ENERGY:
THE CREATED CRISIS

OTHER BOOKS BY
ANTHONY C. SUTTON

WESTERN TECHNOLOGY AND SOVIET ECONOMIC DEVELOPMENT, 1917-1930

WESTERN TECHNOLOGY AND SOVIET ECONOMIC DEVELOPMENT, 1930-1945

WESTERN TECHNOLOGY AND SOVIET ECONOMIC DEVELOPMENT, 1945-1965

NATIONAL SUICIDE: MILITARY AID TO THE SOVIET UNION

WALL STREET AND THE BOLSHEVIK REVOLUTION

WALL STREET AND FDR

WALL STREET AND THE RISE OF HITLER

THE WAR ON GOLD

ENERGY: THE CREATED CRISIS

BY ANTONY C. SUTTON

BRINGING YOU BOOKS THAT MATTER

160 EAST 38 STREET
SUITE 31B
NEW YORK, N.Y. 10016
TELEPHONE: (212) 490-0334

Manufactured in the United States of America

Library of Congress Catalog Card Number 78-73737

ISBN 0-916728-04-8

Cover Illustration: Carlos Mozo-Saravia
Editor: Bill Clark
Index: Paula Nelson
Design: Paulette Nenner

Acknowledgements:
Dobbins' cartoon, reprinted with permission of the *Manchester (N.H.) Union Leader*.

Oliphant cartoon, reprinted with permission of the Los Angeles Times Syndicate

CONTENTS

INTRODUCTION

By U.S. Rep. Steven D. Symms (R-Idaho)

With the imposition of the Arab oil embargo in late 1973, the so-called "energy crisis" was thrust on the American people. We were warned of the impending energy shortage. We were told by those who favor central socio-economic planning that our energy and mineral resources are rapidly running out, that "Spaceship Earth" would no longer be able to support the American people in the manner to which they had become accustomed; and thus, we must reduce our standard of living. "Less is better" became a code phrase for politicians like Gov. Jerry Brown, Congressman Morris Udall, and Jimmy Carter.

The election of James Carter to the Presidency has resulted in much less emphasis on increasing energy production and greater emphasis on government imposed conservation. The growing energy shortage requires "an effective and comprehensive energy policy" which, as the Carter plan decreed, only the government can provide.

However, the doomsayers are wrong. There is *no* shortage of energy *resources*. There is, on the other hand, a shortfall in energy *production*. The *crisis* lies in the fact that the United States is dependent on foreign sources for fully half of its energy and mineral requirements.

In this book, Professor Antony Sutton illustrates our known reserves of energy resources and shows that the United States has within its borders enough fossil fuel energy resources to last us far, far into the future. It is quite clear to the knowledgeable person that, if an energy crisis exists, it exists in other than the category of available energy resources. It is equally clear that the transition of available resources to actually utilizable supply is a function of incentives to accomplish the task.

So the logical question then follows: Why isn't there a greater emphasis on increased energy production?

From my observation, based on six years in the House of Representatives, there seems to be essentially three reasons that government imposed production *disincentives* are not removed, thereby allowing an increase in energy production as the demand arises:

1) There is a phenomenal lack of understanding of basic common sense economics, of the capitalization process and the relationship between personal freedoms and the market economy by the majority of Congress. This problem is especially acute among Congressional staffers who, along with the bureaucrats and other advocates of big government in Washington's numerous think-tanks, are really responsible for most of the socialistic legislation flowing through Congress.

2) Government never gives up power voluntarily. It is truly amazing to observe the lengths to which government and politicians will go to avoid losing even a little power to individuals acting in free and voluntary exchange.

3) There exists a kind of "New Left" coalition of environmental extremists, "no-growth" advocates, and socialistic political groups that has become far more influential in the opinion-making strata of society than their numbers warrant. I have come to discover since being in Congress that the same groups who opposed construction of the Alaska Pipeline co-

ordinated the public campaign against the B-1 bomber: they are now going all out to stop the commercialization of nuclear power.

As of late the godfather of this movement has become one Amory Lovins, an Oxford drop-out who carries the title of British representative of Friends of the Earth. Lovins' initial treatise entitled "Energy Strategy: The Road Not Taken?" was first published by the Council on Foreign Relations in their quarterly journal *Foreign Affairs*. Lovins and his followers advocate the elimination of electric generation systems and distribution networks and replacing them with small backyard systems using only the "soft technologies" such as solar, wind power, etc.

The Congress and the communications media have been flooded by studies (usually funded by the Ford and Rockefeller foundations) advocating the adoption of the Lovins approach as government policy.

A cursory examination reveals the Lovins treatise to be in actuality a formula for de-industrialization; its implementation would be the beginning of America's "Great Leap *Backward*." Advocates of the "soft energy path" and of "no-growth" quite often view these policies as vehicles for social change; they are not talking about alternatives to energy — they are talking about alternatives to a free society. Plainly, what bothers Lovins and his elitist sponsors is not that we have an energy supply problem, but that we might NOT have one.

Antony Sutton illustrates that the so-called energy crisis is a contrived scheme to greatly increase government control over the American people and to further the cause of an egalitarian feudalistic society. As Russell Train, a past Administrator of the Environmental Protection Agency, put it: "We can and should seize upon the energy crisis as a good excuse and great opportunity for making some very fundamental changes that we should be making anyhow for other reasons."

PART ONE

WHAT WE HAVE ...

I

OUR FICTIONAL ENERGY CRISIS

"The basic problem is to turn our whole national policy around — from one that has promoted the widest possible use of energy to one that conserves energy."

S. DAVID FREEMAN, FORMER DIRECTOR, FORD FOUNDATION, ENERGY POLICY PROJECT, NOW IN THE CARTER WHITE HOUSE.

This book has a simple and fundamental message: Our American energy crisis is a counterfeit crisis thrust onto the American people by a politicized elite who have more ambition than common sense.

There is no energy crisis in the sense of a physical scarcity of energy resources. There is, however, a manipulated and artificially created energy situation. There is no absolute shortage of energy resources for our planet in the forseeable future, only a mythological energy Armageddon in the minds of social activists, or a utopian search for a zero-risk world by environmental dreamers.

Our mythical energy shortage can be dismissed with a few statistics. The United States consumes about 71 quads[1] of energy per year. There is available now in the United States, excluding solar sources and without oil and gas imports, about 151,000 quads. Consequently, we have sufficient energy resources to keep us functioning at our present rate of consumption for about 2,000 to 3,000 years — without discovering new reserves. Even at higher consumption rates, there will be no problem in the next millenium.

This book is devoted to explaining this fact — and where we went astray.

The lack of any absolute shortage of energy resources is well known in Washington — although efforts have been made in the Carter Administration to censor release of the documentation. A most revealing picture of our energy abundance was published by ERDA (Energy Research and Development Agency) now part of DOE (Department of Energy). This supply picture is reproduced here as Chart 1-1. U.S. domestic energy consumption in 1977 (71 quads) is represented by the small square in the bottom right corner of the chart. Available energy supplics, with or without new technologies, are represented by the larger squares on the left. The only way to generate an energy crisis is by converting the big potential squares into the little consumption square, and this could only be done by gross and deliberate mismanagement. That is precisely what we have in Washington today.

THE IDEA OF A CREATED CRISIS

The idea of a deliberately created energy crisis is not

(1) A 'quad' is one quadrillion BTUs (i.e., 10^{15} British Thermal Units) — an easily understandable standard measure used wherever possible in this text.

CHART 1-1: ENERGY PRESENTLY AVAILABLE FROM DOMESTIC RESOURCES VERSUS ANNUAL DOMESTIC CONSUMPTION (1977)

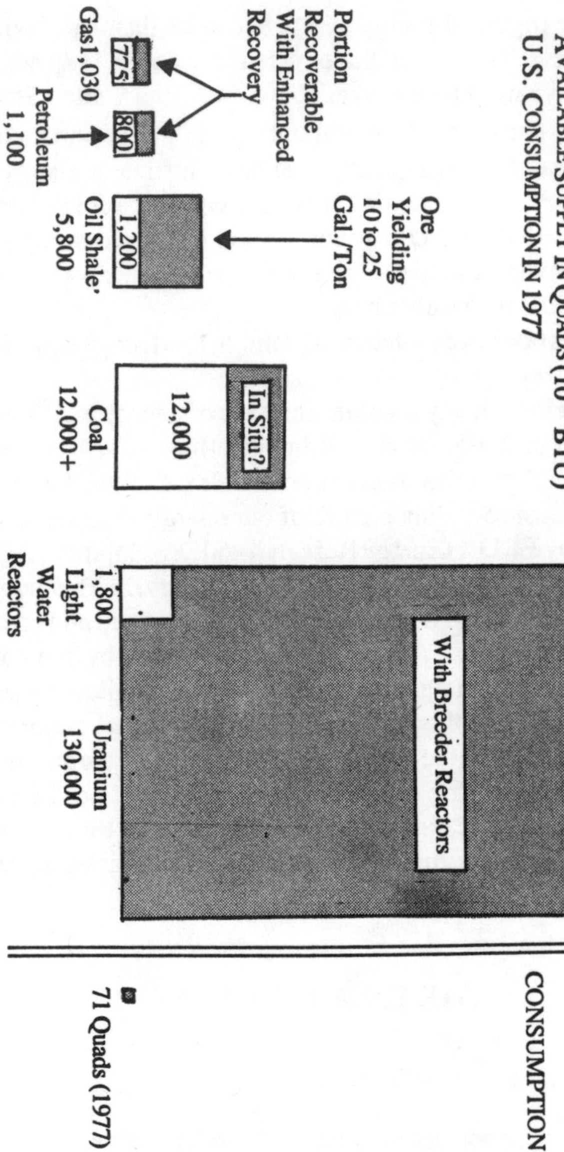

Source: A National Plan For Energy Research, Development & Demonstration, ERDA-48 (1975)

original with this book. The possibility has been advanced by critics of many political and philosophic persuasions who have been forced to the conclusion by the facts.

Back in the "oil crisis of 1973," Governor Kenneth M. Curtis of Maine charged the Nixon Administration "with creating a managed oil shortage crisis to force support of its energy programs." The Democratic Governor accused the Administration of deliberately allowing shortages to reach crisis proportions in order to manipulate the United States into taking certain actions. This accusation had no effect on the Nixon Administration. In January 1974, it used the mythical energy crisis as a scare tactic to pressure oil importing nations to a Washington conference on the grounds that the energy shortage,

"threatens to unleash political and economic forces that could cause severe and irreparable damage to the prosperity and stability of the world."

Was there a contrived shortage of energy in 1973? The evidence appears to be overwhelming. Some commentators who at first hotly denied any such artificial contrivance have, when presented with the evidence, reversed their positions out of professional honesty.

A *PHILADELPHIA INQUIRER* study by reporters Donald Bartlett and James B. Steele[1] in 1973 charged the following:

(1) American multi-national oil firms made deliberate, long-term decisions to expand operations in foreign countries to meet demands for oil products in Europe and Asia;
(2) The Nixon Administration failed to lift oil import restrictions in 1969 and said that there were no oil supply problems;
(3) Simultaneously, American oil companies were telling U.S. cus-

(1) *PHILADELPHIA INQUIRER*, July 22, 1973.

tomers to cut back on consumption, while urging their customers in Europe and Asia to buy more oil products;
(4) The American taxpayer is subsidizing the sale of petroleum abroad through tax allowances and benefits granted to American oil companies;
(5) The gasoline shortage of 1973 was created through default and bungling by oil companies and the federal government;
(6) In 1973, the oil industry launched an advertising campaign to make the American consumer feel responsible for the nation's gasoline shortage. Yet for every barrel (42 gallons) of oil products sold in the United States, the five largest companies (Exxon, Mobil, Texaco, Gulf and Standard Oil of California) sold nearly two barrels abroad;
(7) The percentage of crude oil refined in the United States has steadily declined and has steadily increased in foreign countries;
(8) The demand for crude oil has increased 110 percent in the United States during the past 20 years; in Japan the demand increased 2,567 percent; in West Germany, 1,597 percent and in Italy, 1,079 percent.

Equally weighty refutations come from opposite poles of the American opinion spectrum. The *WALL STREET JOURNAL,* representing orthodox financial opinion, flatly rejects the concept that the world will wake up next Tuesday morning at 7:00 a.m., look out the window and say "My God, we've run out of energy."[1] The *JOURNAL* suggests two ways of looking at the energy problem. The first way is to look at prices, like economists, the second way is to look at stocks of resources, like inventory clerks.

The *JOURNAL* then presents the textbook supply and demand curve equation:

All this means in layman's language is that:

(a) at higher prices people buy less, at lower prices they buy more,
(b) at higher prices more supply is forthcoming, at lower prices less

(1) *WALL STREET JOURNAL*, May 27, 1977.

CHART 1-2: A NO SHORTAGE SOLUTION TO THE ENERGY CRISIS

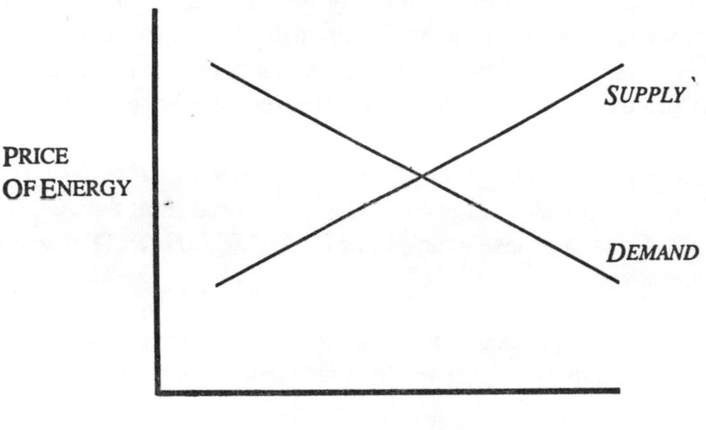

supply is forthcoming,
(c) market price is established where these two forces are in equilibrium.

This is the basic approach used in this book, although we will also explore the "inventory clerk's" approach in some depth.

The *JOURNAL* then claims that the planners and energy strategists, the "zero-growth" Club of Rome, the Carter Administration, the CIA (and it could have added the Ford Foundation, Brookings Institution, and the Congress) are locked onto the idea that "the weight of the earth is finite," i.e., that resources are scarce in some absolute sense. This is an "inventory clerk's view" of the world says the *JOURNAL*, and makes the pertinent point that we are forced to act according to this limited version because of government prohibitions and decrees. In brief, the government creates the energy problem.

Other critics arrive at similar conclusions without the economic analysis. *NEW SOLIDARITY*, at the socialist left end of the political spectrum, in many instances isn't too far away from *WSJ*. Never forget that Lenin instituted the free enterprise New Economic Policy in 1922, and recently the Soviets doubled gas prices to restrict demand in good free enterprise fashion.

The socialist *NEW SOLIDARITY* calls the Carter program "a Nazi energy program," but *NEW SOLIDARITY* recommendations could just as well have come from the *JOURNAL*. Consider these examples of *NEW SOLIDARITY* recommendations:

ON THE BREEDER REACTOR: "commercialize the breeder reactor as quickly as possible." (Note "commercialize," not "government operation");

GASOLINE TAX: "no increase in gasoline tax;"

GAS-GUZZLER TAX: "no gas guzzler tax" (that puts U.S. Labor Party in bed with Henry Ford II);

WELL HEAD TAX: "do not favor well head tax;"

SOLAR ENERGY: "does not support funding for solar and related 'soft' energy forms;"

FOSSIL FUELS: "advocates full exploitation of existing oil and gas reserves."

Source: *NEW SOLIDARITY* "Carter's Nazi Program."

What is the role of the multi-national corporations (MNCs) in the created crisis? Are they on the side of free enterprise and the market solution? Alas, among the multinationals we not only find the articulate proponents of big government but more than a suspicion the MNCs are manipulating big government for their own economic benefit.

Consider the largess distributed as political campaign funds and the "revolving door" influence of the multi-nationals in Washington, D.C. Then look at the extraordinary profits they made after the 1973 "energy crisis." Finally, consider this: In 1977, independent petroleum companies in the U.S. — not the big multi-nationals — drilled 81 percent of all U.S. oil and gas wells and discovered 88 percent of the new oil and gas fields. From this we may deduce that the big multi-nationals are trying to avoid adding to our national supply while they reap the monetary benefits of periodic shortages.

Added to this picture the Carter Administration creation of the mammoth Department of Energy: 2,000 bureaucrats and a $20 billion budget, the seventh largest federal department (in terms of money). To do what? Presumably to ensure that the crisis continues! Almost any issue of a daily newspaper reports the multitude of conflicting administrative actions which keep an energy pot boiling. For instance, on February 2, 1973, the FPC suspended for five months a proposed $42 million rate increase requested by Transcontinental Gas Pipeline of Houston, Texas, thus delaying new gas. On the same day, New York restricted use of natural gas in certain end uses because of a "shortage." Also on that same day, thousands of New Yorkers shivered from lack of fuel oil while officials refused to allow heating oil to be shipped in — the sulfur content was too high.

We opened this chapter with an epigraph, a quotation from S. David Freeman, who today oversees the created crisis from the Carter White House. We can aptly conclude this chapter with another Freeman quotation:

"The energy crisis is a self-inflicted wound. It is not Mother Nature, but Uncle Sam that is to blame."

This then is the topic of *Energy: The Created Crisis*. We approach the energy crisis essentially from the economist's viewpoint, but let's first put on our "inventory clerk's" hat and summarize the stocks of energy resources available to us using existing technology or technology we can reasonably expect to be developed given market incentives.

Chapter II

II

AMERICA'S ABUNDANCE OF COAL

"We are running out of power and have no plans that can close the energy gap in time to avert a shutdown of American industry, and with that, a collapse of our way of life and loss of our national independence."

LAWRENCE ROCKS AND RICHARD P. RUNYON, *THE ENERGY CRISIS*, (NEW YORK: CROWN, 1972) P. xiv.

The United States has vast, almost inexhaustible reserves of coal. Approximately one-third, some sources say one-half, of the world's recoverable coal reserves, i.e., coal that can be recovered using present technologies under existing conditions, is located in mainland United States. These bountiful reserves now available amount to at least 200 billion tons, of which 45 billion are near the surface and the balance more deeply underground, distributed fairly evenly among the eastern Appalachian region, the mid-West and Western states.

Is there a shortage of coal? There is no shortage of coal, here or anywhere in the world. We can expand use of coal and

TABLE 2-1: WORLD COAL RESOURCES
(recoverable reserves)

	IN BILLION TONS	
United States	200.3	30.8
Soviet Union	150.5	23.1
Europe	139.7	21.4
China	88.1	13.5
Oceania	27.0	4.2
Africa	17.2	2.6
Rest of world	28.9	4.4
Total	651.7	100.0

Source: World Energy Conference, *Survey of Energy Resources*, 1974.

still find ourselves with ample reserves. If we have an energy crisis, it's not because we are running out of coal, now, in the next century or in the next millenium.

UNIVERSAL AGREEMENT ON COAL ABUNDANCE

Maximum possible production from now until 1985 would use only about ten percent of the coal, assuming no further reserves will be discovered and added to the already considerable total. Counting coal reserves not presently utilized, in the words of Herbert Foster, vice president of the National Coal Association:

"America has 3 trillion tons of coal out there, ready to be mined.
"And all we produced last year was 590 million tons. That's only one pound of coal for every 2½ tons still in the ground!"

Even the U.S. Department of the Interior admits the United States has reserves sufficient until well into the next century and the Energy Research and Development Administration, now part of the Department of Energy, has stated:

"The United States has vast resources of coal. Our domestic coal reserves are equal to one-half of the known reserves in the entire world and have five times the energy value of our domestic recoverable oil and natural gas."

Lawrence Rocks and Richard Runyon — who initiated the energy crisis scare with their book *The Energy Crisis* — agree that "Our coal energy bank is theoretically large enough for at least a thousand years assuming the present rate of consumption will remain constant," and add that if coal is used to produce synthetic gas, this use will reduce coal reserves by only about two centuries. A Rocks-Runyon inconsistency should be noted: in their opening pages, the lifespan of United States coal reserves is cited as "200 to 300 years if coal is used to synthesize oil and gas at their present growth rates."[1] Later in the book, they insert the more realistic span of 1,000 years.

In fact, the total coal reserves of all types known to exist in the United States have been estimated by the U.S. Geological Survey (more acceptable than any other source) at more like 3.2 trillion tons.[2] In other words, the oft quoted 200 billion ton figure of recoverable reserves may be less than five percent

(1) Compare pages 9 and 33 of Lawrence Rocks and Richard Runyon, *The Energy Crisis*, (New York: Crown, 1972).
(2) Energy Research and Development Administration, *Coal In Our Energy Future*, (Washington, 1977).

of potential U.S. domestic coal supplies. What this adds up to is more than enough coal in the United States to supply all comceivable domestic uses and expanding exports well into the coming centuries, which allows ample time to develop new technologies and methods for coal reserves not presently mined. Given the distribution of world coal reserves, as shown in Table 2-1, it is quite possible that the U.S. could become a major coal exporter.

THE ROAD BLOCK TO DEVELOPMENT

What's holding up development of coal reserves?

Unfortunately, 40 percent of U.S. reserves are beneath land held by the Federal Government — and the U.S. Department of Energy, as we shall see later, is the multi-billion dollar, 20,000 employee roadblock to energy production. In addition, ecologists, on the assumption that wildlife is a superior life form to humans, are actively delaying all forms of energy development.

Moreover, everything has a cost, including coal mining. The cost of mining coal will undoubtedly rise during the coming decades due to price inflation, less favorable geological conditions and health and safety measures under the 1969 Coal Mine Health and Safety Act; these sum to a probable increase of one-third in coal prices during the next decade. While a rising cost of coal may be safely predicted, what cannot be predicted accurately is relative costs of producing energy from different sources during the coming years. Thus, no valid economic decision can be made favoring one form of energy development over another — and this decision is precisely what so many people are trying to do. Only the market place, with a minimum of government intervention, will give us an efficient energy mix.

At the end of the Nineteenth Century, coal supplied about 90 percent of the U.S. energy consumption. During the first decades of the Twentieth Century, coal was largely replaced by natural gas and oil by the operation of the market place because gas and oil were cleaner, more convenient and more competitively priced than coal. By 1950, coal provided only 38 percent of our energy. However, this substitution process was carried much too far because the government intervened into the market process and accelerated replacement of coal by artificial stimulation of nuclear energy, and the later implementation of the Clean Air Act which required many coal users to convert to oil. So by 1972, only 17 percent of energy came from coal.

Finally, there is not too much controversy among those with access to geological fact: coal is abundant, there is no shortage, we have all the coal we can conceivably use in the next few thousand years. However, natural gas is a different case and involves greater disagreement among observers. This suggests we should take a closer and detailed look at natural gas reserves.

Chapter III

III

HOW MUCH NATURAL GAS DO WE HAVE?

"We're not running out of gas. We're running out of cheap, readily available gas. That's an important distinction."

ROBERT E. SEYMOUR, CHAIRMAN, AMERICAN GAS ASSOCIATION

Natural gas, one time unwanted orphan of the petroleum industry and vented or flared off in the oil fields, has emerged today as our cleanest, cheapest, and most convenient fuel. In many end uses, gas can be substituted for coal, which has expensive pollution drawbacks. Gas even has been used in Europe as a substitute for automobile gasoline. Widespread domestic use of natural gas for cooking and home heating makes it of immediate concern for householders. Most of this natural gas comes from wells operating along the Texas-Louisiana Gulf Coast and transported by pipeline to consumers in the contiguous United States through a $50 billion, 1 million mile transmission and distribution network. Current annual domestic demand for natural gas is 20 quads,[1] of which about 8.5 quads goes into industrial use and 7.6 quads for domestic use, the remaining 3.9 quads is for electricity generation and miscellaneous end uses.

For almost a century, commentators have asserted that natural gas is almost exhausted, that we are running out of a precious asset, and a crisis is both inevitable and imminent. In 1893, more than three-quarters of a century ago, Karl Baedeker provided the following information for European visitors to America:

"The store of rock gases (natural gas) known to exist in this country will probably be exhausted within twenty years of the present time. The resources in the way of petroleum are also likely to be used before the middle of the next century. The fuel in the form of coal exists in such quantity that there is no reason to apprehend a serious diminution of the store for many centuries or perhaps even thousands of years to come."[2]

We got past 1913 safely without the gas wells becoming exhausted. Under President Truman, just a quarter of a century ago in June 1952, the Paley Commission (The President's Materials Policy Commission) again reported pessimistically about future supplies of natural gas. The Paley Commission believed that gas supplies had actually limited demand, even at higher prices, while natural gas resources were being used up at an alarming rate. Consequently, the Commission inserted in its report a section entitled "Preparing for eventual transition" to

(1) One quad = 1 trillion cubic feet (TCF) of natural gas.

(2) Karl Baedeker, *The United States: A Handbook for Travellers - 1893*, (New York: Da Capo Press reprint, 1971) p. lxxix.

plan for a time when other sources of fuel would be substituted for gas.[1]

In 1972, twenty years after the scary Paley prognosis, the Bureau of Natural Gas (BNG) of the Federal Power Commission published a report on the anticipated supply and demand balance for the years 1971 to 1990. The Bureau was as pessimistic as Karl Baedeker in 1893 and the Paley Commission in 1952:

"Our 20-year forecast to 1990 indicates that the rate of development of natural gas supplies both conventional and supplemental, willl be inadequate to meet current projections of future demand."[2]

The Bureau of Natural Gas envisaged consumption of gas falling further behind theoretical demand, with annual supply deficits of 9 TCF by 1980 and 17 TFC by 1990. BNG estimated production would peak in the mid-1970s, then dip, with imports filling 40 percent of consumption by 1990. The dip in reserves is to be sizeable in the contiguous 48 states, "dropping from its present level of 259.6 to 170.4 trillion cubic feet by 1990."[3]

To arrive at this dire prediction, the Bureau of Natural Gas, the government's gas expert, ignored new technology and brushed aside unconventional sources of gas, just as the Paley Report ignored unconventional technologies. Supply for BNG will come only from existing and proven reserves which ob-

(1) The President's Materials Policy Commission, *Resources for Freedom: Foundations for Growth and Security,* (Washington, D.C.: February 1972) Bureau of Natural Gas, Staff Report No. 2, p.1.

(2) Federal Power Commission, *Natural Gas Supply and Demand 1971-1990,* (Washington, D.C.: February 1972), Bureau of Natural Gas, Staff Report No. 2, p.1.

(3) *Ibid.*

viously are in continuing decline, some Alaskan gas, liquid natural gas imports and limited coal gasification. Consequently, the BNG estimates are absurdly low. For example, coal gasification is put at 3.3 TCF by 1990, well below what is technically feasible, even with existing technology. Supplemental sources, i.e., liquified natural gas and coal gasification can easily yield 6 quads by 1990. In brief, while BNG concludes we shall have a total 186 TCF cumulative deficit by 1990, it may be what the White House wants, but is remote from geological and technical reality.

This gloomy outlook was also reflected in the report of a committee established by the Secretary of the Interior and published as *U.S. Energy Outlook*[1] in December, 1972. The report examined future production under several hypothesized supply conditions. The basic assumption, "The Initial Appraisal," was that the 1970 pattern of government regulation, policy, and economic climate will continue into the 1971-1987 period.[2] Then the committee assumed that its "Initial Appraisal" was too optimistic "and that significant changes were essential if the adverse trend in energy supply is to be improved." So four hypothetical cases were postulated under varying conditions of incentives, government regulation and drilling rates. By comparing estimated 1975 production for these four cases to actual results in 1975 we note that the 1972 "Initial Appraisal" natural gas supply was 19.8 TCF, while actual domestic production in 1975 was 19.9 TCF — which is close enough. However all the hypothesized cases (I through IV) suggesting production of between 21 and 23.7 TCF were proven wrong by actual 1975 production. If the Committee had

(2) *Ibid*, p.1.

stayed with its interim, preliminary appraisal and make no further assumptions, much of the study would have been unnecessary. This further confirms the point that without an energy crisis there would be no need and no work for energy "experts."

These official prognostications of doom have been reflected by the media and even by a few industrial spokesmen. In September 1972, *Fortune* climbed onto the energy crisis bandwagon, with an article "The energy 'joyride' is over" and commented on natural gas as follows:

"Our cleanest, most convenient fuel — natural gas — may be the first to give out. Leaving aside the Alaskan discoveries the U.S. since 1968 has been using natural gas twice as fast as it has been finding it. In many parts of the country gas companies are refusing to make gas available to heat new homes, and are forcing some industrial users to shift back to oil when home heating demand is high."

Among the industrial Jeremiahs, C. Howard Hardesty, executive vice president of Continental Oil Company, added his warning in a speech May 18, 1972.

"Natural gas is already in critically short supply. Domestic production of gas is projected to decline from almost 22 trillion cubic feet in 1970 to 14.5 trillion cubic feet in 1985..."[1]

Almost every major U.S. publication picked up a theme of a dire natural gas shortage, a necessity to cut back in life styles and modify the standard of living, a need to tap solar energy, stop washing in hot water, and so on. It appears that the less knowledge there is of the facts, the more extreme becomes the proposed solution.

(1) Speech to Public Utilities Association of the Virginia News Seminar, Pipestem, West Virginia, May 18, 1972.

So let's take a look at the known and potential gas reserves of the United States, published and readily available to government officials, politicians, writers, editors and preachers of natural gas crises.

RESERVES AND PRODUCTION OF NATURAL GAS

To establish the absence of any real shortage or crisis in natural gas, we have to identify two sets of statistics:

a) What is the annual production of natural gas? and
b) what is the extent of our natural gas reserves from which this production is drawn?

A century ago there was infintesimal production of natural gas in the United States. In 1900, natural gas production was .254 TCF or 3.2 percent of all energy fuels used.

By 1920, this minute gas production had quadrupled to .883 TCF and still was about the same percentage (4.1 percent) of total energy production. During the 1920s production of natural gas soared, reaching 2.1 TCF by 1930 (or 9.7 percent of all energy production). The depression of the 1930s restrained demand and in 1940 only 2.9 TCF was used (11.9 percent of all energy production). After World War II there were more increases in production from 4.4 TCF in 1945 to 6.8 TCF in 1950, 14.1 TCF in 1960, and 24.2 TCF in 1970. Comprising 13.7 percent of energy produced in 1945, the natural gas share climbed to 34 percent in 1960 and 38.7 percent in 1970. By 1975 the percentage declined slightly to 36.9 percent.

For much of the last half century, recoverable gas reserves have increased roughly along with consumption and production of gas. In 1925 reserves were estimated at 23 TCF compared to production of 1.3 TCF; in 1941 the Petroleum

CHART 3-1: U.S. NATURAL GAS RESERVES (1947-1974)

(in trillion cubic feet)

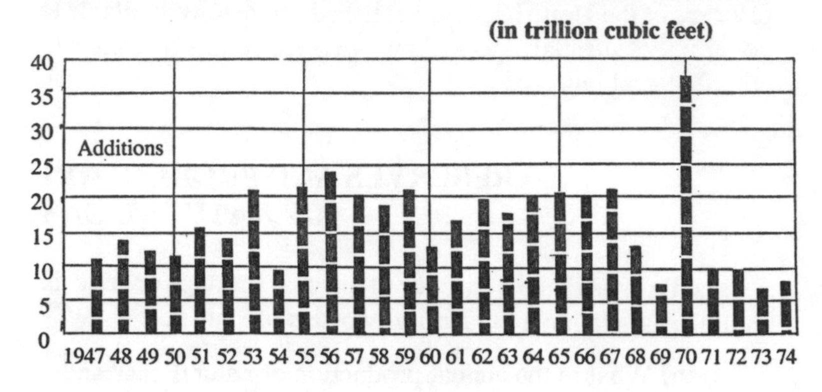

CHART 3-2: U.S. NATURAL GAS PRODUCTION (1947-1974)

(in trillion cubic feet)

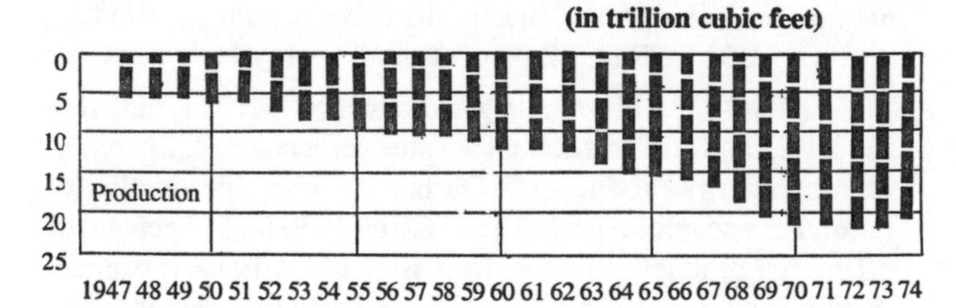

Source: American Gas Association and American Petroleum Institute.

Administration for War estimated reserves at 113.8 TCF with production of 3.2 TCF; in 1945 reserves estimated by the American Gas Association as 147.8 TCF compared to production of 4.4 TCF. In the early years, gas production was only a twentieth or a thirtieth of known gas reserves. Thereafter the reserve-production ratio shifted unfavorably.

TABLE 3-3: NATURAL GAS IN THE UNITED STATES — ANNUAL PRODUCTION AND CONSUMPTION AS PERCENTAGE OF PROVEN RESERVES (in trillion cubic feet)

	1950	1955	1960	1965	1970	1975
Total Production	6.8	10.5	14.1	17.7	24.2	22.2
Total Consumption	6.1	9.2	12.7	16.1	22.0	20.2
Proved Reserves	184.5	223.6	263.7	286.4	290.7	208.7
Annual Consumption as % of Proven Reserves	3.3%	4.0%	4.7%	5.6%	7.6%	12.5%

Source: American Gas Association

From 1950 to the present day, both production and consumption (production plus imports minus exports) have increased in relation to proven reserves. From 3.3 percent in 1950 the production-reserve ratio shifted to 12.5 percent in 1975. The reasons for this uncomfortable shift will be explored in Chapter Seven ('Why Can't We Have Natural Gas?')

We need first to take a look at the total reserve of natural gas available. Far from a shortage, we find that our abundance of gas is almost embarrassing: to match the current consumption of 20 to 25 TCF a year, we can count on total reserves — from all sources — in excess of 3,700 TCF. That's about a century or more of usage at current rates, which is fair enough ahead for most of us.

When the *WALL STREET JOURNAL* editorialized the news that we possess "1001 years of natural gas" DOE Secretary James Schlesinger accused the *JOURNAL* editors of "smoking pot." As the facts are so clearly on the side of the

WALL STREET JOURNAL in this instance, we begin to see emerge a created crisis by Schlesinger and his friends. What are these facts?

As of October 1973, only about one percent of the Outer Continental Shelf had been leased and, today, perhaps two percent, despite strong evidence that the shelf contains more than one-half of our potential gas and oil resources. Further, more than one-third of the potential gas supply of the contiguous United States is located in southern Louisiana and the Texas Gulf. More than half of our current domestic gas production comes from these two prolific areas, while along the Atlantic offshore areas, there is a potential of 12 billion barrels of oil and 67 TCF of gas. Yet at the time of this writing less than a dozen exploratory offshore wells have been drilled between Maine and Florida.

Why? That interesting question we leave to Chapter Seven.

A similar set of evidence comes from Herman Kahn and the Hudson Institute:

> "Allowing for the growth of energy demand estimated earlier, we conclude that the *proven reserves* of these five major fossil fuels (oil, natural gas, coal, shale oil and tar sands) alone could provide the world's total energy requirements for about 100 years, and only one-fifth of the *estimated potential reserves sources* could provide for more than 200 years of the projected energy needs!"[1]
>
> (Exclamation point and italics in original)

This natural gas abundance is distorted or ignored by Establishment elements intent on proving an energy crisis. For instance, President Carter's program arbitrarily eliminates about one-third of our natural gas capabilities. Table 3-4

(1) Herman Kahn, et al., *The Next 200 Years*, (New York: William Morrow, 1976), pp. 63-4.

illustrates the gas industry's estimate of its 1985 supply capabilities, based on existing plants and contracts, compared to the Carter Administration plan:

TABLE 3-4: 1985 NATURAL GAS SUPPLY PROJECTIONS (ANNUAL QUADS — 10^{15} BTUs PER YEAR)

SUPPLY	NATURAL GAS INDUSTRY CAPABILITY	THE CARTER PROGRAM	ADDITIONAL CAPABILITY BEYOND CARTER'S PROGRAM
Continental U.S. Natural Gas	20.0 quads	17.0 quads	3.0 quads
SNG from Petroleum Feedstocks	1.2	0.5	0.7
Coal Gasification	0.4	—	0.4
Alaska Gas	1.2	0.1	1.1
LNG Imports	2.0	0.6	1.4
Canadian Imports	0.6	0.6	—
Total	25.4	18.8	6.6

Source: American Gas Association, *Gas Supply Review*, September, 1977.

How much natural gas do we have? Estimates of remaining conventional resources calculated by various authorities are contained in Table 3-5.

These estimates begin at low industry figures in the 700 to 800 TCF range to the 1,000 TCF of the U.S. Geological Survey. The figures do not include unconventional gas re-

TABLE 3-5: NATURAL GAS REMAINING IN THE UNITED STATES (TCF in 1974) From Conventional Sources Only

SOURCE OF ESTIMATE	YEAR END	NEW FIELDS	OLD FIELDS	PROVED RESOURCES	TOTAL
United States Geological Survey	1974	322-655	202	237	761-1094
National Academy of Sciences	1974	530	118	237	885
Moody	1974	485	65	237	787
Garret	1974	500	100	237	837
Exxon	1974	582	111	237	660-1380
Mobil	1973	443	65	250	758
Potential Gas Committee	1972	880	266	266	1412

serves because sufficient demand does not yet exist for serious development of these sources. However, to complete the picture, four such unconventional sources (not included in Table 3-5) are: (1) coalbeds, (2) shale formations, (3) "tight sand" formations, and (4) deep underground water zones in the northern Gulf of Mexico basin.

Methane, the prime constituent of natural gas, is a by-product of coal formations, and has been mined commercially in small quantities in the eastern United States since about 1940. Most coal mine gas is presently vented — to the order of about 75 BCF per year, and it is estimated that at least 240 to 300 TCF of coal gas is available in the contiguous United States. Development problems include the low concentration of the gas and its low permeability, although fracturing techniques can increase gas flow and permeability. Draining this gas from coal mines does have an important value. It increases safety because methane is a dangerous mining hazard. Although a most promising long-term supplementary source of gasification, production and exploration have been restrained for decades by the artificially low prices for natural gas imposed by state and federal regulatory commissions.

Sometime around 1971, the Washington bureaucrats in the BNC became aware of an impending gas shortage, and a shortage is the only signal which will galvanize the planners into new technology. Private industry was not paying attention because the industry's Bituminous Coal Research, Inc., labs worked closely with government and looked forward to government grants for research. Private research has lagged waiting for federal funds. The industry has been unwilling to invest its own capital because it knows that government funds will be forthcoming as shortages became apparent to slow-moving politicians and bureaucrats.

After the 'Clean Air Message' of June 4, 1971, the American Gas Association and Department of the Interior finally funded an eight-year, $296 million program for coal gasification research. Three pilot programs for coal gasification research. Three pilot programs were started: the Institute of Gas Technology HYGAS plant in Chicago, the Consolidated Coal Company's CO_2 Acceptor Process in Rapid City, South Dakota and a plant at Homer City, Pennsylvania, using the BI-GAS process developed by Bituminous Coal Research, Inc. That's where we are today, gradually moving into coal gasification at the rate set by slow moving planners rather than the more rapid rate of the profit-oriented free market.

A second unconventional source, not presently developed, is the Rocky Mountain "tight sands" formations estimated to contain 600 TCF; i.e., a reserve equal to all known conventional gas reserves. "Tight sands" have less than normal space between grains; this gives them low permeability, and causes difficulty in extraction of contained gas. One method of loosening up these sands was attempted in Operation Plowshare, a federal government financed program of underground nuclear explosions designed to fracture the sands and release the gas. Operation Plowshare included Project Gas Buggy, Project Rulison and Project Blanco. The nuclear explosion of Project Gas Buggy was denotated December 10, 1967, at 4,200 feet underground in a joint experiment by the Atomic Energy Commission, the U.S. Bureau of Mines and the El Paso Natural Gas Company. Subsequent production tests yielded more than 100 million cubic feet of gas, compared to a conventional well (400 feet away) which required nine years to produce 85 million cubic feet. Project Rulison exploded a 40 kiloton atomic device at Battlement Mesa, Colorado. The site was sealed for six months to allow heat to dissipate and then opened for production. On May 17, 1973, under Project Rio Blanco, three nuclear devices were denotated under Piceance Creek basin in Colorado.

A third unconventional source of methane is shale, underlying much of the United States between the Appalachians and the Rocky Mountains — an area of about 250,000 square miles. These shales are fine-grained rock of little pore space and low permeability. In the past fifty years, gas wells have been drilled in one section of this potential producing area. Explosives are used to fracture the shale and improve permeability.

In 1972, Columbia Gas Systems initiated a program to develop information on the best way to recover natural gas from the shale deposits. Columbia Gas Systems has two contracts with ERDA for hydraulic fracturing testing. To fracture layers several thousand feet thick, fluid and sand are increased many times to give a more powerful hydraulic fracturing called massive hydraulic fracturing (MHF). The first Columbia Gas Systems contract called for four MHF treatments in three shale wells in Lincoln County, West Virginia. The second contract involved 14 MHF tests in 13 other wells. Five out of six wells receiving MHF treatment have increased production significantly.

ERDA also sponsored a five-year program to assess gas reserves in the entire 250,000 square mile area of shales. This program should increase reserves equal to several times the estimated proved reserves in the United States.

There is a fourth source of methane not included in present estimates and is known as "dissolved-in-water" methane. The solubility of methane in water is a function of pressure, water temperature, and salinity. In the northern Gulf of Mexico basin between depths of approximately 10,000 and 15,000 feet, water trapped underground reaches temperatures as high as 520°F and pressures as high as 15,000 pounds per square inch. Energy recovered from this resource could well exceed domestic proved reserves. As the methane is dissolved in saline water, it may be necessary to pump enormous quantities of water from underground to obtain gas, that is, to produce about five percent of present annual production, it may be necessary to drill approximately 1,000 wells, each well costing as much as $3 million. One geologist suggests that as water is pumped out, pressure on the remaining water underground will be reduced, and so permit some gas to move out of the water and enter porous rock, in time creating new reservoirs of gas. This idea still has to be confirmed.

Finally, another unconventional, but very practical and renewable source of natural gas is to grow methane in giant seaweed marine farms. California Giant Kelp *(Macrocystis pyrifera)* grows at a speedy two feet per day. Plants are anchored in large plots 60 feet deep in the ocean; the rapidly growing plants near the surface are lopped by mechanical harvesters, and chemical processing of the kelp cuttings yield methane, fertilizer and carbon dioxide. This process is technically feasible and has already been tested in small offshore farms. It now remains to test commercial production to determine the relative economics of the biomass system. In brief, the United States can ultimately be self-supporting in natural gas by converting to marine biomass production after exhaustion of the 2,300 TCF conventional resources.

With no imports and no marine origin gas, the United States has at least sufficient natural gas for more than a century at present rates of consumption. Reasonable imports, with or without marine origin gas, can cover extra demand for economic growth. This bountiful reserve does not include extensive underdeveloped resources overseas. Mexico will be cited in Chapter Seven. Australia is investing $3.3 billion to develop its Northwest shelf gas reserve; the gas will be liquified and exported at an annual rate of seven million tons over 20 years, to the United States — if we want it.

In addition, there are readily available supplemental sources. The quickest is a process by which light gasoline fractions are processed into synthetic gas (SNG); synthetic gas plants came into operation in 1973 and today there are 13 operating SNG plants producing about 1 TCF per year. SNG is a private response to a demand situation. It is private initiative using private capital and not the result of any government scare crash program approach. In fact, federal authorities have currently suspended an additional 5 SNG projects.

Other projects envisaged use of naphtha and propane as feedstocks for synthetic gas production. Such projects are presently limited by oil import control regulation. Liquid gasification can provide synthetic gas volume of 1.5 trillion by 1980, but requires a new view of oil import controls to allow foreign feedstock to supplement inadequate domestic light fractions. Those projects which do not require federal regulation nor depend upon imported feedstock are moving ahead. An even quicker source of imported supplemental gas is pipeline import from already producing wells in western Canada; (1 TCF was imported from Canada in 1972). Large gas discoveries in Arctic Canada can be linked by pipeline to the Midwest and West Coast, and by 1990, annual Canadian gas imports can reach four trillion cubic feet. We can also look to Alaska for gas via a trans-Canadian pipeline. North Slope gas is associated with oil and cannot be recovered until the oil is produced. It is also possible to bring in about ten percent of U.S. natural gas consumption in the form of liquified natural gas (LNG) from Algeria by about 1985. While Algeria is not a large domestic consumer today (making it a willing exporter) its developing economy will create domestic demand by the 1990s. The most advanced major project for importing LNG is that of El Paso Natural Gas, to move one billion cubic feet per day of natural gas equivalent in liquid state from Algeria to two terminals on the East Coast of the United States. The investment for this undertaking is estimated at nearly $1.8 billion. Other gas utilities have announced similar projects. The LNG supplemental supply is expected to increase to 3.5 TCF per year by 1990.

MINIMUM RESERVES 3760 TRILLION CUBIC FEET

So when we add up our natural gas reserves, we find a bountiful supply indeed.

While natural gas resources in the ground are more than ample — 3,760 TCF will keep us going for another century or more and transmission and distribution facilities are fully able to take care of peak demands, although claimed shortages of natural gas began to appear in 1972 and have continued well into the decade.

TABLE 3-6: UNITED STATES RESERVES OF NATURAL GAS FROM ALL SOURCES (in trillion cubic feet)

		TCF
CONVENTIONAL SOURCES	Proven Domestic Gas Reserves	237
	Undiscovered gas (NSF-1975)	530
UNCONVEN-TIONAL SOURCES	Tight sand reservoirs	600
	Coal associated gas (current technology only)[1]	250
	Shale formations	500
	Underground water zones in Gulf of Mexico	200
	SNG from peat [2]	1443
RENEWABLE SOURCES	Marine bio-mass	
TOTAL RESERVES AVAILABLE		3,760 TCF

Sources: American Gas Association, U.S. Energy Research and Development Administration.

Note: (1) Total estimates for coal associated gas range up to 794 trillion and estimates for undiscovered gas range over 1,000 TCF.
(2) AGA *Gas Supply Review*, December 1977.

Chapter IV

IV

A CENTURY OF OIL AHEAD

"The President (Carter) said there is no chance of us becoming independent in our oil supplies. That is just wrong. We have at least 100 years of petroleum resources in this country."

WILLIAM BROWN, DIRECTOR OF TECHNOLOGICAL STUDIES, HUDSON INSTITUTE.

Remember the spring and fall of 1973? The gasoline shortage; the snaking lines of cars at the gas pumps? The rush to the stores to buy five-gallon cans; 55-gallon drums and lock caps for gas tanks? We certainly had a gasoline crisis, with allocations, gasoline rationing expected daily and even the millions of necessary coupons printed up by an ever watchful Washington.

Then the crisis was all over as quickly as it began. Gas stations re-opened. Refinery output crept back up to the normal 50 million barrels a week to equal average weekly consumption. It had seemed serious enough, but when it was all over, the American Automobile Association was unable to find a single case of a motorist stranded for lack of gasoline — even in refinery short areas such as New England and upstate New York. The real difference after the crisis was the price at the gas pump. Prices climbed to the 60-cent range, and gasoline at the 40-cent range became history.

A long known axiom was highlighted by the 1973 gasoline shortage — one can always trust a government to exacerbate a crisis: price ceilings were introduced at the peak of the scarcity. What do government price ceilings do? They restrict supply and prolong the scarcity. The quickest way to relieve any shortage is to let prices go where they want. The extra profit brings forth further supply and prices generally come back down, more or less, to earlier levels.

We shall also see that, up to now, there has been a notable lack of success in gaining political objectives — whether OPEC embargoes or a Carter tax plan — with oil and gasoline. Not only is surplus production capacity available but there is an elastic and flexible world petroleum transportation system and a world oil market. Merely by increasing production in countries with surplus capacity, re-scheduling tankers and drawing down inventories — and it does this much more quickly, if governments leave prices alone — the petroleum industry has always delivered in time of crisis and avoided gasoline rationing. On the other hand, the United States is potentially highly vulnerable to crude oil embargoes by virtue of its extraordinary dependence on the imported hydrocarbon.

THE AUTOMOBILE SOCIETY

Automobile-oriented America is critically dependent on petroleum products. Furthermore, the percentage of these petroleum products produced from domestic oil reserves has been steadily shrinking. In 1950 domestic production of crude petroleum was 1,974 million barrels. In 1974, domestic production was 3,203 million barrels — an increase of 60 percent. However, imports of crude petroleum between 1950 and 1974 soared seven-fold from 178 million barrels in 1950 to 1,269 million barrels in 1974. In brief, while only eight percent of petroleum was imported in 1950, 28 percent was imported in 1974. This does not allow for exports, which are relatively small and declining: 35 million barrels in 1950 and one million barrels in 1974.

CHART 4-1: SOURCES OF U.S. ENERGY, 1975

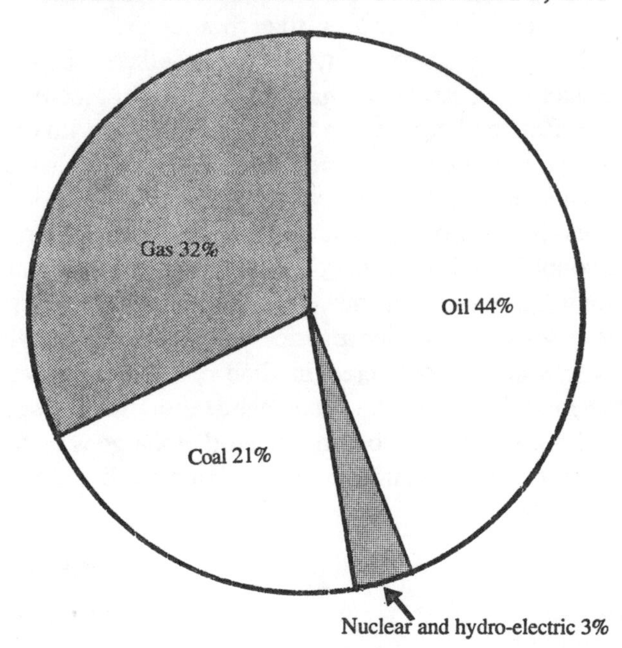

Coupled with dependence on imported oil, we can see from Chart 4-1 that oil is the largest portion of our total energy sources: in fact, crude oil is at the moment the most important single energy source for the United States: 44 percent compared to 32 percent for gas and 21 percent for coal. Moreover, the fuel balance of the U.S. is quite different from that of the Soviet Union and Europe, making this a uniquely American problem. While the United States is heavily dependent on petroleum whether its source is domestic or foreign, the Soviet Union is by no means as significantly dependent on petroleum as a fuel:

TABLE 4-1: PETROLEUM AS A PERCENTAGE OF ALL ENERGY SOURCES

UNITED STATES	(crude oil and natural gas)	76 percent
EUROPE	" " " "	56 percent
SOVIET UNION	" " " "	36 percent

Source: American Petroleum Institute

The American automobile, a vitally important contribution to the economy, is the basic reason for these proportions.

The Soviet Union derives only 15 percent of its energy from liquid petroleum while the greater part is derived from coal (57 percent) and natural gas (21 percent). In some measure this reflects the comparative underdevelopment of the present Soviet economy; for example, the stock of motor vehicles in the Soviet Union is only about 1/25th that of the United States, and is well below that of European countries.

In the United States on the other hand, the internal combustion and the diesel engine are key prime movers; railroads have been converted to diesel and ocean going ships are largely propelled by marine diesels; aircraft use fuel refined from petroleum almost exclusively.

TABLE 4-2: PASSENGER CARS PER 1000 POPULATION (1975)

UNITED STATES	497
U.S.S.R.	19
WEST GERMANY	294
ITALY	270
UNITED KINGDOM	251

Source: *World Motor Vehicle Data*, Motor Vehicle Manufacturers Association of the U.S., Inc.

The Soviet Union will not face the automobile fuel dilemma until at least the 1980s since it has only 19 autos per 1000 population compared to 497 in the U.S. Then, with its gigantic reserves, the U.S.S.R. will face a production problem requiring vast imports of Western technology and major capital investment.

ARE WE RUNNING OUT OF CRUDE OIL?

Given this rate of consumption, and our reliance on imports and the automobile, are we running out of crude oil?

Whether we have a shortage of crude reserves can be answered initially as follows. In 1976, the United States produced 3.6 billion barrels of crude oil and natural gas liquids, while the proven reserves of crude oil at the end of 1976 were 37 billion barrels, i.e., about ten years at the current usage rates. In addition, the estimated recoverable reserves, under current economic conditions, were 150 billion barrels or about 40 years at current usage. The total cumulative United States crude production, from 1859 through 1976 was about 115 billion barrels, so we have at least more recoverable crude reserves remaining than have been drawn upon in all production to date.[1]

What is the difference between proven and recoverable reserves? Proven reserves consist of crude oil recoverable from existing known oil reservoirs (the 37 billion barrel figure includes the North Shore Alaskan field). However, estimated recoverable reserves refer to reservoirs which are known to exist geologically but have not been developed sufficiently to arrive at a precise engineering estimate of recoverable reserves. Numerous pessimistic estimates of U.S. reserves have been made since the turn of the century, and turned out to be false because they ignored identified reservoirs without development work and with unknown reserves.

To be realistic, when we discuss running out of crude oil, we must include presently inferred reserves where scientific support can be found for their probable existence. In other words, we have to estimate unknown reservoirs of oil. If we do not make this estimate, then our statements become absurd. We are not running out of oil if we take into account the likelihood of further discoveries under normal economic and business conditions.

These unmeasured reservoirs can be approximated by looking at offshore sedimentary basins. The offshore fields are almost all underdeveloped: all drilling so far has been in waters

(1) These figures can be calculated in various ways. For example, the American Petroleum Institute's *Annual Reserve Report at January 1, 1977*, places currently recoverable crude at 30.9 billion barrels and crude oil currently unrecoverable under today's technology and economics as 303.5 billion barrels, i.e., 80 years supply.

less than 600 feet deep, and out of 28 major offshore fields, only two, the Central field in the Gulf of Mexico and the Santa Barbara Channel in California, have been brought to production. Few have been developed beyond the initial evaluation stages.

About two percent of the continental shelf has been opened up by the federal government to leasing. Continental shelf wells are currently producing only about 615 million barrels of oil and 3.8 trillion cubic feet of natural gas annually. These unknown reserves are in four regions: (see map) beneath the Atlantic Ocean, beneath the Gulf of Mexico, beneath the Pacific Ocean and under Alaskan waters.

U.S. Marine Producing and Potential Petroleum Areas

The total crude oil potential of this outer continental shelf has been estimated by the U.S. Geological Survey.

CONTINENTAL SHELF AREA	ORIGINAL OIL IN PLACE[1] (to 25,000 meters)
ATLANTIC OCEAN	224 billion barrels
GULF OF MEXICO	575 " "
PACIFIC COAST	275 " "
ALASKA	502 " "
TOTAL	1,576 billion barrels

Source: Based on U.S. Geological Survey study by McKelvey, March 11, 1968.

Another vast untapped oil reservoir consists of the oil shale beds in Wyoming, Utah and Colorado. Thousands of square miles already appraised contain oil shales more than ten feet thick and yielding 25 or more gallons per ton of shale. Additional thousands of square miles contain either low yield shale formations or are unappraised. Certainly there are numerous technical and environmental problems with oil shales, but nothing insurmountable. The oil is there. What is needed is development work, the application of technological skill.

Looking at the favorable oil geology picture in the U.S. even more broadly, the National Petroleum Council estimates more than three million square miles of the United States contain geological formations favorable to oil and gas accumulation. That's about four million cubic miles of potentially favorable sediments. Production is currently under way in about 50,000 square miles, i.e., less than two percent of the total favorable area. On a global basis, some 500 or so sedimentary basins are known, of which three basins probably

(1) "Based on estimated oil in place per unit area in selected U.S. land areas, as applied to analogous offshore areas."

contain more than one half the world's known oil reserves: the Persian Gulf, Western Siberia and the Gulf of Mexico.

Finally, still looking at the wider picture, one of the most reliable overall petroleum reserve estimates comes from former director of the U.S. Geological Survey, V. E. McKelvey:

"...at this point about half the world's endowment of presently recoverable crude oil has been discovered (and) about one-sixth of it has been used up."[1]

THE NATURE OF THE OIL CRISIS

Given the estimate that only one-sixth of our oil potential has been used, then clearly we have no fundamental oil crisis on our hands for the next few decades. Our current problem is to develop still bountiful crude oil reserves. Uncertainties only can be removed by getting down to work, with more exploration, more drilling and application of new recovery methods. Uncertainties are not removed, but are compounded, by political bargaining and planning strategies.

The key factor is that oil reserves are not static and that proved reserves increase according to economic and operating conditions and particularly financial incentives. These factors are closely linked to energy prices, technology and available substitutes. As the American Petroleum Institute has commented:

"Proved reserves at any time are quite distinct from forecasts of what will be made into reserves later. Recent discoveries in northern

(1) *BARRONS*, November 20, 1977.

Alaska were originally said to 'contain' from 5 to 10 billion barrels. This was the amount expected to be developed in the course of years. Eight months later there was an implicit reserve forecast, when it was announced that a pipeline built from the area would have a capacity of 2 million barrels daily (MBD), which indicates about 15 billion barrels...."

The mistake we have made is to rely on government bureaucrats and planners to solve a problem they cannot solve because they are not in a position to make the calculated risk decisions involved in oil exploration. Moreover, their own jobs depend on the existence of a continuing problem. There is no incentive for government planners to find a solution, and as we observe daily, the emphasis of the government approach to the so-called energy crisis is restriction of consumption, not expansion of supply.

We shall explore government-created problems later. Let us first run down the real difficulties that do exist in fact, have always existed, but can be overcome by a capable petroleum industry. First, there is competition for labor, material and capital; oil competes with other energy projects and with non-energy goods and services for resources. This has always been the case. And the impartial market place, left alone, does an excellent job of allocating resources to their best uses.

Second, there is technology. Expansion of crude oil production in the U.S. needs utilization of secondary and tertiary recovery. Much of the technology has not been used commercially, but the principles are generally known and can be applied if the incentives exist.

Third, there is the problem of access to petroleum resources. The federal government controls 40 percent of the remaining recoverable oil, and makes this land available for exploration and development only with extraordinary delay and hesitation. The terms under which government land is available affects availability of capital, the rate at which leased areas are explored and the percent of oil-in-place ultimately recovered. Further, additional state regulations also influence production rates, recovery and costs.

Fourth, government prices and regulatory policies, depletion allowances, tax rates, and fiscal uncertainty all constrain investment and so production.

Fifth, there are environmental considerations. All extraction, manufacturing and distribution processes affect the quality of the environment — and oil is no exception. Oil spills affect the marine environment and can create aesthetic problems. This we discuss in Chapter Eight.

All these are real, but not insurmountable difficulties, and they do not contain the crux of the petroleum dilemma. The crux of our dilemma is in the nature of government intervention into the oil picture. At this point, let us confine ourselves to reserve and production statistics as calculated and propagandized by energy bureaucrats. The argument supports the claim that bureaucrats and politicians can only ensure that an energy crisis continues. It is the incentive for them to find and perpetuate crises. The following chart, by way of example, is taken from the Energy Research and Development Administration and illustrates two key statistics: (a) "cumulative production through 1974" and (b) "remaining recoverable" oil which can be produced through 2020.

The ERDA chart is misleading. There was no factual basis in 1975 for assuming that U.S. domestic production had peaked in the early 1970s, to be followed by a steep decline, unless this decline is to be deliberately brought about by government policy. We have seen that oil reserves are known and in place, sufficient for at least 50 years. Yet, the ERDA chart reports 182 billion barrels total remaining recoverable

CHART 4-2: U.S. PRODUCTION OF PETROLEUM LIQUIDS

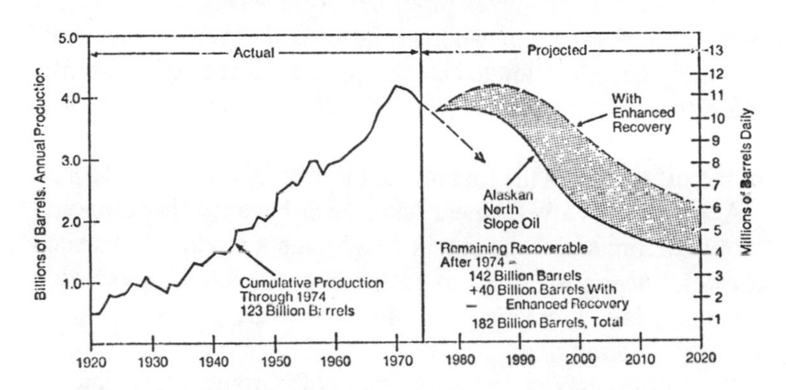

Source: *A National Plan for Energy Research, Development & Demonstration*, ERDA-48 (1975).

reserves at a time when offshore in-place original oil alone has been estimated at 1,576 billion barrels — more than eight times the ERDA figure. In brief, the chart creates the illusion of a shortage where none exists.

Similarly, in the Project Independence reports there is parallel distortion of data. One example will suffice to make the point. A project report [1] makes the following statement:

"At $11 per barrel oil, domestic onshore production would increase slightly under both BAU and AD assumptions. Almost half of the onshore production could be from new secondary and tertiary recovery, while conventional and new primary fields would decline considerably from 1974 levels."

NOTE: BAU is "Business As Usual" and AD is "Accelerated Development."

(1) Federal Energy Administration, *Project Independence Report,* PB-248-492, November 1974.

What is the "Accelerated Development" (AD) scenario? "AD" allows further government leases for offshore oil exploration, legalizes pipelines from Alaska and releases water and environmental constraints on oil shale development. In brief, the "Accelerated Development" scenario is an artificial situation where the government stops behaving in a manner which impedes oil production. If the government had kept out of the oil picture in the first place, there would be no need for an "Accelerated Development" scenario because the economy would automatically be in that development mode. Under such artificial scenarios, we can justifiably call the oil shortage a created crisis, a creature of political power, or if you want it bluntly — a phoney.

Let us take one last example of statistical misbehavior to press the point home, specifically the potential oil production figures released by the old Federal Energy Administration.

Table 4-3 (Columns 1, 2 and 3) is reproduced from the Project Independence Report and again estimates potential oil production assuming $11 a barrel under the business constraints "BAU" (Business As Usual) and "AD" (Accelerated Development). In the 1970s there was a gap between production and consumption of about six billion barrels per day (6.4 in 1973). This gap will be almost closed (except for expansion in consumption) by 1985 under the Business As Usual scenario and closed with additional production for growth under Accelerated Development. However, these successes, which are considered to be only achievable by 1985, can be realized long before that time, even in the 1970s. Oil from Naval Petroleum Reserve #1 and #4 (listed in the FEA table at 2.2 billion barrels a day) is ready for almost instant exploitation. Production from conventional fields can be expanded very rapidly simply by removing state quotas and price controls. Alaska and the North Slope was held up for years by government red

TABLE 4-3: POTENTIAL RATES OF DOMESTIC OIL PRODUCTION (millions of barrels per day, at $11 oil)

PRODUCTION AREA	1974	BAU 1985	AD 1985	Production 1973	Consumption	Gap
Offshore - Lower 48 states	8.9	9.1	9.9			
"Conventional fields and new primary fields	6.9	3.4	3.5			
"New secondary	—	2.4	2.4			
"New tertiary	—	1.8	2.3			
"Natural gas liquids	2.0	1.5	1.6			
"Naval Petroleum Reserve #1	—	—	0.2			
Alaska	0.2	3.0	5.3			
"North Slope	—	2.5	2.5			
"Southern Alaska (including OCS)	0.2	0.5	0.8			
"Naval Petroleum Reserve #4	—	—	2.0			
Lower 48 Outer Continental Shelf	1.4	2.6	4.3			
"Gulf of Mexico	1.3	2.1	2.5			
"California OCS	0.1	0.5	1.3			
"Atlantic OCS	—	—	0.5			
Heavy Crude Oil & Tar Sands	—	0.3	0.5			
Total Potential Production	10.5	15.0	20.0	10.9	17.3	6.4

Source: Federal Energy Administration, *Project Independence Report,* PB-248-492, November 1974.

tape. Tar sands projects have an off-and-on history of government decision and indecision. In sum, Table 4-3 distorts reality because under a simple scenario never examined, i.e., "Government Removes All Restrictions" both the BAU and AD options can be attained promptly. Of course, if we adopt the most straightforward and obviously successful scenario, we have no need of Washington experts to devise energy strategies and scenarios.

RED FACES AT THE ENERGY DEPARTMENT

Those who watch governments and politicians solve problems by creating more problems were not surprised when, in 1977, far from the predicted running out of oil, the world experienced a surplus of oil, a glut. The 1973 experts who worked up the crisis scenarios had forgotten to include new oil from the North Sea and the Alaskan slope into their equations. So in 1977-78 there was a world surplus of about 3 to 4 million barrels a day. And while the North Sea and Alaska expanded production, the old time suppliers in Saudi Arabia, Venezuela and Libya reduced output. Saudi Arabia imposed a maximum crude production for 1978 of 8.5 million barrels a day on Aramco, which produces most Saudi oil. However, because of the glut Aramco has been unable to reach even this limit, selling only 7.2 to 7.6 million barrels a day in early 1978. These receding oil markets cut Nigerian production from 2.2 million barrels a day in 1976 to 1.6 million barrels a day in 1977 and to 1.2 million barrels a day in 1978 — and that's a traumatic production cut for a lesser-developed country in economic trouble. In Venezuela the planners budgeted for 2.2 million barrels a day, but production in February 1978 averaged only 1.5 million barrels a day.

What happened? Simply oil markets at work, a process neither understood nor welcomed by the energy planners. Rising oil prices in 1973-74 encouraged the search for oil, i.e., a normal response to incentives. From 1974 to 1977 newly found oil reserves exceeded production by five billion barrels a year instead of the officially projected decline. Mexico entered the league of big oil producers and by 1977 its proven reserves were 14 billion barrels, about 40 percent of the United States' reserves. North Sea oil turned Britain's economic decline around in a dramatic manner, reducing inflation and encouraging production. If oil proceeds are invested in technology, Britain may withstand economic storms until atomic energy can take up the burden at the turn of the century. New geological discoveries increased U.S. reserves to the 1.5 trillion barrel figure, enough for 65 years or so given existing consumption rates. And of course higher prices encouraged exploratory drilling, and exploratory drilling leads to new discoveries. It is this drilling rate, which requires capital and incentives, that dominates new oil supplies. If the finding rate (barrels per foot drilled) holds up, more exploratory and developmental drilling will mean more oil and gas. During the last decade or so, the drilling and the finding rate declined. The oil drilling rate declined from about 200 million feet per year in the 1950s to about 100 million feet per year in the 1970s, and at the same time the finding rate declined from more than 400 barrels per exploratory foot drilled to less than 100 barrels. Oil production declined because we were not finding oil and bringing it to the surface. Given the incentive of a higher price for crude, the drilling rate increased, although the finding rate must decline in the long term for obvious geological reasons. Wells have to be drilled deeper.

Contrast this optimistic, realistic scenario to the 1974 Project Independence summary report which concluded: "Between 1974 and 1977 there is little that can prevent domestic production from declining or at best remaining constant."[1]

(1) *Project Independence: Summary Report,* p.5

Today energy officials are wiggling their charts and statistics a little harder to locate reasons to convert a glut into a shortage. The oil surplus situation cannot last is the current theme. True, markets operate to even out surpluses and — if governments keep out of the marketplace — will balance demand and supply. We may have temporary changes but still need to hold the fundamentals in mind. As Alice M. Rivin, Director, Congressional Budget Office stated:

"A major reason for the substantial increase in our import dependence is the current system of price controls on oil and gas that have kept the domestic price of these fuels artificially below world levels...It has also tended to encourage energy consumption and discourage the search for and production of new domestic resources — thereby further increasing our dependence on potentially unreliable foreign suppliers."

Unfortunately, we now have a gigantic, expensive bureaucracy dependent on a crisis. By arguing that shortages will return by 1981 and 1985, they will keep the multi-billion dollar, 20,000-employee Department of Energy political machine in business.

Chapter V

V

AN ATOMIC CORNUCOPIA

"We can juggle growth curves and cross our fingers, but it is becoming abundantly clear that nuclear power, in the 'last resort' will be crucial to our well-being."

CARL WALSKE, PRESIDENT,
ATOMIC INDUSTRIAL FORUM.

We have three potential sources of almost inexhaustible energy, in addition to the currently used atomic fission reactor: the breeder reactor, fusion power and solar energy. Breeder reactors are currently in operation in several countries. Fusion power will be unfeasible for many decades because of tough engineering tasks yet to be solved. Solar energy has usefulness under certain economic and climatic conditions but suffers from an insoluble defect: it is a diffuse source of energy and requires a great deal of land surface for solar collectors. Physical laws make solar energy of limited usefulness for large scale energy generation under many, if not most, environmental conditions.[1]

NUCLEAR POWER IS CHEAPER

There are presently 70 nuclear power reactors operating in the United States, with a total generating capacity of 50,431 megawatts electrical (MWe). Another 90 units have construction permits (97,719 MWe), four have limited work authorizations (4,626 MWe), 47 units are on order (53,253 MWe), and one has a letter of intent (1,150 MWe), for a total of 207,179 MWe operating or on the way. In 1976 nuclear energy provided 12 percent more power than in 1975 while oil and coal generation increased by 11 percent each.

Are these nuclear plants safe? Consider the following: The United States has approximately 2000 reactor-years experience of military and commercial reactor operations without a single accident with injury to a member of the public. Approximately 300 reactor years of civilian reactor operations are included, while commercial light-water power reactor designs used in the United States have accumulated approximately 175 reactor-years of operation without a single reactor-caused injury. Moreover, nuclear plant availability has been almost as good as oil and coal and at a lower unit cost:

To quote the *WASHINGTON POST*, our choice for the future seems to lie between coal and atomic energy — but coal

(1) See Petr Beckmann, *The Health Hazards of Not Going Nuclear*, (The Golem Press, Box 1342, Boulder, Colorado 80306), and compare to D. Hayes, *Energy: The Solar Prospect*, (Washington, D.C.: Worldwatch Institute).

is certainly more dangerous to produce and may be more expensive.

"To generate increased amounts of electricity, this country now has only two choices. It can either burn more coal or build more uranium-fueled nuclear reactors. Sometime in the next century other technologies will emerge — perhaps solar or geothermal generators. But for the next 30 years or so, this country must expect to depend on the sources that we already have; it's necessary to weigh these two fuels, coal and uranium, against each other. What are the respective risks for health, safety and the natural environment? Coal, on present evidence, is more dangerous than the present generation of nuclear reactors running on enriched uranium. Coal is also likely to be a bit more expensive than nuclear power in most parts of the country."

	AVAILABILITY* FACTOR		FORCED* OUTAGE RATE	
	1976	1975	1976	1975
Nuclear Plants	71.6%	73.8%	14.2%	13.7%
Oil Plants	75.5%	78.9%	14.2	13.7%
Coal Plants	75.9%	78.6%	11.1%	11.1%

* Availability is a measure of the time the plant is in good operating condition.
★ Forced outage is a measure of unnecessary shut-down time.

Looking even further into the future, the breeder reactor is by far the most promising of the long range alternatives. The breeder uses U_{238}, in addition to more scarce U_{235}, and converts U_{238} to plutonium. One pound of plutonium provides the same energy as three million pounds of coal. The significance is that a breeder reactor uses about two tons of uranium oxide per 1,000 megawatts per year, i.e., about 1/79th that of conventional water cooled reactors. Furthermore, not only does the breeder economize existing uranium resources by stretching current known U.S. uranium resources to about

4,500 years, but breeders can use the waste depleted uranium from uranium enrichment plants. In the United Kingdom, enough such depleted fuel is already available to meet U.K. energy needs for 100 years, equal to all presently known British coal reserves. In the United States 150,000 tons of depleted uranium is available at ERDA uranium enrichment plants. Used as fuel, this depleted uranium would generate as much electricity as 400 billion tons of coal — a little less than one-half of our coal reserves.

Uranium can only be used practically to fuel nuclear powered electricity generating stations. A study by the Council on Economic Priorities (CEP) has concluded that coal would have a cost advantage over nuclear generation in the United States everywhere except in the North East where the two systems would be on parity. This was not a study of actual costs, however, but a study of projected costs based on certain assumptions. A later study by Perl rejected many of those assumptions as unrealistic. For example, CEP gave too much importance to older nuclear reactors which are more expensive to run, and made assumptions about future costs not borne out in practice. When actual nuclear production costs are compared to costs for coal and oil generation, nuclear power has a decided advantage. In fact, this is what we would expect. Utilities would not undertake the capital burden of nuclear generation unless there was a clear cost advantage. Table 5-1 summarizes these cost advantages for many existing nuclear, oil and coal units presently at work.

Some of these generating stations use all three methods of electricity generation. For example, Virginia Electric & Power has 23.3 percent nuclear capacity. The cost of its oil generated power is 2.7¢ per Kwh, coal 1.8¢ and nuclear 1.3¢. The overall weighted average figures for all stations listed in Table 5-1 is oil 3.5¢, coal 1.8¢ and nuclear 1.5¢ per Kwh.

TABLE 5-1: ELECTRICAL GENERATING COSTS IN THE UNITED STATES, 1976 (by source of energy)

UTILITIES	NUCLEAR PORTION OF CAPACITY	TOTAL COST PER Kwh (in cents) NUCLEAR	OIL	COAL
Arkansas P & L	42.3%	1.5	2.5	*
Atlantic City Electric	18.0	1.6	3.4	1.9
Bangor Hydro Electric	43.2%	1.0	3.4	*
Boston Edison	22.4%	2.9	3.3	*
Central Maine Power	45.0	1.0	3.2	*
Commonwealth Edison	40.0%	1.4	4.1	2.0
Consolidated Edison	8.1%	3.8	4.0	*
Dairyland Power Coop	5.5%	2.2	8.4	1.5
Duke Power	25.0%	1.2	*	1.7
Eastern Utilities Assoc.	17.5%	1.8	2.8	*
Green Mountain Power	59.2%	1.7	3.7	2.1
Jersey Central	55.8%	1.5	5.1	—
Metropolitan Edison	29.7%	2.9	8.9	2.0
Nebraska Pub. Power Dist.	69.4%	1.6	1.6	1.4
New England Electric System	14.1	1.2	—	*
New England G&E System	26.1%	1.7	2.9	*
Niagara Mohawk Power	10.0%	1.4	3.2	2.1
Northeast Utilities	56.0%	1.4	3.2	*
Omaha Pub. Power Dist.	42.8%	1.2	5.8	1.2
Philadelphia Electric	25.8%	1.6	4.6	2.1
Power Authority, N.Y.	23.3	1.3	*	*
Pub. Service Co. of New Hampshire	12.6%	1.1	—	2.2
Rochester G&E	32.0%	1.5	*	—
Sacramento M.U. Dist.	67.7%	1.2	*	*
Southern Calif. Edison	4.1%	0.8	—	0.8
Virginia Electric and Power	21.6%	1.3	2.7	1.8
Wisconsin Electric Power	43.0%	1.0	*	1.9
Wisconsin Pub. Service	23.3%	1.8	—	1.9
Weighted averages [1]		1.5¢	3.5¢	1.8¢

Source: Atomic Industrial Forum.

INEXHAUSTIBLE CHEAP ENERGY

Opponents of nuclear power in the United States have spread the idea that breeder reactors are an unknown and distant technology, probably impractical or at least highly dangerous. By contrast to this propaganda, Great Britain, France, West Germany and Japan all have thriving advanced breeder projects in operation. And, as will be seen in Chart 5-1, these programs are considerably ahead of the United States, where President Carter has been evading the issue for political reasons.

In April 1977, the West German cabinet proposed a $2.7 billion, four-year program for energy research. A sizeable proportion of this funding is scheduled for fast breeder reactors and to reprocess depleted uranium fuel. Also in April 1977, Japan started up an experimental fast breeder reactor, JOYO, at a cost of $100 million. In May 1977, Premier Raymond Barre of France made the decision to go ahead with SUPER PHENIX, a French second stage breeder reactor: (the original 250 MW PHENIX was operating in 1974). An ambitious breeder reactor program is under way in Great Britain with a prototype 250 MW breeder reactor put into operation during 1974 and a start made on a commercial size 1300 MWe breeder in 1977.

By contrast, the U.S. approach is exemplified by the *Ford-Mitre Study, Nuclear Power Issues and Choices*.[1] This concludes that the U.S. must use light water reactor technology

continued from previous page

(1) Weighted average total cost per Kwh for coal and oil combined: 2.2 cents. Fuel Cost 1,3 cents.

— = Not available.

* = Not applicable

CHART 5-1: DEVELOPMENT OF BREEDER REACTORS

	UNITED KINGDOM	UNITED STATES	FRANCE
1954	ZEPHYR test reactor		
1955			
1956			
1957			
1958			
1959	DOUNREAY (DFR)		
1960	60 MW Reactor		
1961			
1962			
1963		ENRICO FERMI plant	
1964		demonstrator	
1965			
1966			
1967			
1968		continuous problems	
1969			
1970			
1971			
1972		ENRICO FERMI plant	
1973	closed		
1974	Prototype fast reactor		PHENIX 250 MW
1975	250 MW demonstrator		in operation
1976			
1977	Start construction com-	Clinch River demon-	Premier Barra approves
1978	mercial size reactor	strator reactor delayed	SUPER-PHENIX
1979		by President Carter	
1980			Planned Commercial
1981			PHENIX generation
1982			
1983		CLINÇH RIVER 350 MW schedlued for	

continued from previous page

(1) Report of the Nuclear Energy Policy Group, *Nuclear Power Issues and Choices,* (Cambridge, Mass., Ballinger Publishing Co., 1977).

(LWR) for the next half-century, and defer using breeder technology. This is supposedly required by the global proliferation of nuclear weapons and the assumption that deferring breeder development will not be harmful to our energy situation. There are differences of opinion whether a breeder will add to nuclear proliferation, and the cost of delaying or bypassing the breeder are much larger than the Ford-Mitre Study anticipates.

The proliferation argument ignores the "Cive" process known to the Ford-Mitre researchers. This process obtains plutonium fróm nuclear waste but not at extract weapons grade. The fuel remains highly radioactive throughout the production process — a useful deterrent to would-be thieves.

Breeder benefits are large even with the assumptions of the Ford-Mitre Study [1] and, of all advanced sources of power, the fast breeder is closest to realization. Its scientific feasibility was demonstrated as far back as 1942 when Enrico Fermi achieved the first sustained fission reaction. Experimental breeder reactors have been operating since the 1950s. But engineering problems are formidable since the fast breeder reactor uses liquid sodium as a coolant to transfer heat from the reactor core to a steam-producing boiler.

But it seems extraordinary that we should delay development of such promising technology. That this delay comes from an Establishment think-tank (Ford-Mitre) is also interesting in view of the source of the funding for the anti-nuclear campaign.

(1) Rene H. Males, and Richard G. Richels, *Economic Value of the Breeder Technology:* A comment on the Ford-Mitre Study, April 12, 1977.

ASSAULT ON NUCLEAR ENERGY

Nuclear power plant construction has been the target of a numerically small, well-funded, sometimes quite effective political protest movement intent on halting power plant construction. Backing up these on-site protests is a flood of antinuclear pamphlets and materials, in significant part funded by Establishment foundations such as the Union of Concerned Scientists Fund, Inc., which receives funding from the Rockefeller Brothers Fund. The Worldwatch Institute (anti-nuclear) received $500,000 from the same source.

Two of the most publicized on-site protests have been Diablo Canyon in California and the Seabrook Plant in New Hampshire. In a real sense, these are artificial media-created events. Without press coverage, protests are not news. To twist a saying from the Viet Nam War, "what if they held a protest and the press didn't come?" Moreover, in the case of Diablo Canyon, reports suggest a cozy agreement between the Abalone Alliance protestors and the local police department to generate a scenario for the event. The protesters would illegally trespass but be non-violent about it and the police in turn would arrest the non-violent protesters and be gentle about the arrest process.[1]

All this was photographed, reported and editorialized for public consumption. An editorial in the August 2, 1977 issue of the San Luis Obispo *Telegram-Tribune* claimed the newspaper did not intend "to be used by the Abalone Alliance or any other group," but subsequently, ample reports were

(1) See reports in the San Luis Obispo *Telegram-Tribune*, July and August 1977.

printed and numerous pictures run on the protest activities of the Alliance.

Diablo Canyon nuclear plant is owned by Pacific Gas and Electric Company. The company first applied for an operating license in 1973. In August 1977, PG&E still awaited grant of an interim license in the face of persistent and noisy invasions of the property. Unit One of Diablo Canyon (1 million KW) is ready for its interim license. The cost of the delay, including interest charges and additional replacement fuel oil, has been $15 million a month — all ultimately paid for by California power consumers.[1]

What is the problem at Diablo Canyon?

A much-publicized problem was a crack in a water pipe that fed the steam generator. A faulty weld was discovered during normal test operations. It was a routine fault commonly encountered in new facilities. But, because this was a nuclear plant, the weld is good for 30 column inches in the local newspaper[2] conveniently printed just at the time when the Abalone Alliance was assembling for a noisy invasion of the plant. This episode was followed on August 4, 1977, by an editorial charging "slip shod workmanship" and "more questions than answers at Diablo," based on charges by the anti-nuclear forces without a company response.

The August 1977 protests designed to halt the grant of an interim license were given front page coverage by the San Luis Obispo *Telegram-Tribune*. Subsequently, Congressman Panetta, who is too smart a Congressman to be sucked into a con game like this, was given front page prominence in another local newspaper denouncing "U.S. Nuclear Policy,"[3] and opposing the breeder reactor because of higher costs and

(1) Letter from Pacific Gas and Electric, August 9, 1977

(2) San Luis Obispo. *Telegram-Tribune*, July 23, 1977.

(3) *Sun Bulletin*, Morro Bay, September 29, 1977.

nuclear proliferation. Yet buried within these attacks, overwhelmed by the noise of irrelevant argument, is some common sense. Panetta attacked the "special interests" influencing Congress, while Barry Commoner, another anti-nuclear publicist, wanted more emphasis to be placed on solar power and less on nuclear power because the latter will "produce a technical base for fascism." This argument reminds us of Lenin's plans for electrification of the Soviet Union as the basis for socializing the State, and there is little question that central power systems can be used as a collectivist political control device.

A favorite anti-nuclear environmental argument is "thermal pollution," that is, the discharge of atomic power stations warms sea waters with disadvantageous effects on marine life. This argument is silly. We find, in practice, that marine life flourishes under "thermal pollution." The kelp beds off Encina Power Plant, in San Diego County are a favorite haunt of unusual fishes never seen as far north as San Diego: the *sectator ocyrus* and the *kyphosus analogus* have been caught here, and nowhere else off California. The Cultured Catfish Company of Colorado City, Texas, reports its catfish grow much faster in the warm waters off a Texas Electric Service plant. The Long Island Oyster Farms, Inc., report that oysters grow five times faster in the warm waters off the Long Island plant at North Port, New York.

The beneficial effects of raising the ocean temperature have implications for another nuclear hassle: the trials and tribulations of the Seabrook Nuclear Power station, a $2 billion unit designed to supply 80 percent of New Hampshire's power in the 1980s — if Seabrook ever gets into operation.

Seabrook has been subject to a most extraordinary series of regulatory indecisions, inefficiencies, and inconsistencies summed up well by a *Wall Street Journal* headline "Beserk proceduralism." Seabrook has been attacked in and out of court by a cluster of anti-nuclear lobbies, generating more heat than light, protesting, marching, sign-waving and generally creating a fear atmosphere around nuclear construction. Of the 12 years it will need to build Seabrook, five years have been taken up with regulatory indecisions and court proceedings. Almost 50 different permits had to be fought for from the Federal Nuclear Regulatory Administration, the Army Corps of Engineers, the State of New Hampshire and local authorities. Seabrook is absolute proof that a government's function is always to restrict. Government has no constructive capability.

The Public Service Company sums it up:

"Seabrook is caught in a classic Catch 22. The facility is needed. It is a non-petroleum burning source of electric energy. Those state and federal authorities with siting authority (NHSEC and NRC) agree it should be built at the chosen site. But two federal agencies, each of which has power to enforce its will, have been unable to settle upon the type of cooling system to be used. While EPA has the power to dictate the type of cooling system, NRC has the power to reject the entire project because of its view that the EPA decision incurs other environmental (e.g., air and land) costs which result in the overall costs outweighing the benefits."[1]

In one extraordinary series of decisions, a local EPA regional administrator approved the seawater cooling system which would raise the temperature of the sea offshore about one quarter mile away by about five degrees. As noted above, this is certainly not harmful, and may be beneficial to marine life. This decision required 110 days. The Nuclear Regulatory

(1) Public Service Company of New Hampshire, News Release, April 19, 1977.

Commission issued a permit. Subsequently, an EPA regional administrator changed his mind and stopped construction. An incoming EPA administrator, Douglas M. Costle, decided it all required a second look. He appointed a panel of six EPA scientists, and upon their report approved the project. At this point, the Seacoast Anti-Pollution League went back into court and began another round of delays, where we are today. All because a few clam larvae might get sucked into the Seabrook seawater cooling system!

Similarly, the Consumers Power nuclear plant at Midland, Michigan, is now eight years behind schedule and five times over an original cost estimate of $350 million. Why? Because an unreasoning group of anti-nuclear protesters has intervened in the construction process and used the regulatory process to delay construction. It is agreed that in this case a Chicago attorney, Myrin Cherry, is the culprit, "He has bullied, badgered and outraged a small legion of adversaries," according to the *Wall Street Journal* (March 10, 1978).

In effect, Congress has legislated the right for a small and unrepresentative group of individuals to burden the economy with vast costs, expensive power and probably in some cases, no power at all. It is a situation similar to the transfer of American technology to aid Soviet military power, while at the same time we spend $100 billion a year or so on defense against the Soviet Union. The United Kingdom and France are developing and building commercial breeder reactors while the U.S. delays and restricts development at the highest levels of government. Even in light water reactors using a time-tested technology, the U.S. delays by government regulation. Even though no one has been killed in a nuclear reactor accident (and hundreds are killed each year in coal mine accidents) we insist on creating a fear atmosphere around atomic energy. Do we find demonstrators protesting coal mine accidents? Do we find

Congress willing to repeal anti-business legislation and allow power plant construction to go ahead? Do we find a White House willing to follow Europe's technical lead in breeder reactors? The answers to these questions suggest why we have an energy crisis.

THE WHITE HOUSE
WASHINGTON
August 19, 1977

Dear Mr. Sutton:

I have been asked to respond to your letter of July 28.

Beginning in 1949, during the early period of atomic submarine development, President Carter worked with a small group of Naval officers who had a joint responsibility to the submarine force and to the Atomic Energy Commission.

He was senior officer of the crew of the Sea Wolf, teaching mathematics, physics and reactor technology. The President was involved in helping construct a prototype power plant at Knolls Atomic Power Laboratory.

As part of his preparation for this work, President Carter and other officers studied special graduate courses in reactor technology and nuclear physics. The President's undergraduate degree is a B.S. in Engineering.

We hope this information will be helpful.

Sincerely,

Landon Kite

Landon Kite
Staff Assistant

VI

AN OVERVIEW OF OUR ENERGY INVENTORY

"For the most part, Department of Energy's mission, through no fault of its own, is to prevent the U.S. energy industry from discovering, producing and distributing energy."

WALL STREET JOURNAL,
JANUARY 17, 1978

The preceding four chapters have summarized the energy reserves of the United States. Obviously there is no absolute shortage of energy raw materials. Clearly, we are not going to run out of coal, gas or oil next Monday morning at 10 o'clock, or next year, or in the next century. But the energy crisis debate does not concern itself with the amount of energy resources available. It is conducted on the totally erroneous assumption that resources are virtually finished and crisis is now upon us. The theme of the energy debate simply has been who is to control the future use and mix of energy. It is a debate over future control, not future supply.

Before we explore this observation at length, let us tally the energy resources available to us and get an overview of our abundance.

TABLE 6-1: OVERVIEW OF AMERICA'S ENERGY RESOURCES

ENERGY SOURCE	KNOWN DOMESTIC RESERVES (conservative estimates)	TEXT REFERENCE BELOW:
Coal	200 billion tons	Chapter 2
Natural Gas	3760 trillion cubic feet	Chapter 3
Crude oil	1.5 trillion barrels	Chapter 4
Atomic energy	Almost unlimited using breeder reactor	Chapter 5

While absolute reserves are plentiful, a significant adverse factor is increasing scarcity of currently mineable grades. As deposits are mined, ore grade declines. This is a fact of economic geology. Consequently, costs of mining rise and so the final price of energy must rise. Any attempt to restrict a price rise due to declining grade under the guise of "helping" the consumer will merely generate shortages. It is critical to understand that we cannot avoid the geological facts of life. Someone, somewhere is going to have to pay the price of increasing resource scarcity.

Another important factor is that consumption is rising because of both economic growth and population increases. How many years of each energy resource do we have at present consumption rates? Table 6-2 explains.

These resource estimates are conservative. For example, where Table 6-2 lists 200 years of natural gas, the *Wall Street Journal* has identified "1001 years of natural gas," and a Bureau of Natural Gas estimate (in 1972 before Schlesinger

TABLE 6-2: YEARS OF DOMESTIC ENERGY AVAILABLE AT PRESENT RATES OF CONSUMPTION

	QUADS	YEARS SUPPLY AVAIL–ABLE AT 80 QUADS CON–SUMPTION ANNUALLY
Natural gas	2,300	200 years
Petroleum	1,100	130 years
Oil shale	5,800	1,500 years
Coal	12,000	6,000 years
Atomic (light water reactor)	1,800	600 years of uranium U_{238} in storage for 100 years of breeder use.
Atomic (breeder reactor)	130,000	

Source: ERDA: See Chart 1-1.

took office) is 340 years. In fact, as we have seen using biomass methods, we could list natural gas as unlimited in supply. Thus, ew have at least 2,000 years of fossil fuels, and 2,000 years is as far in the future as most of us need to look.

Official and semi-official energy studies reflecting this geological reality are hard to find, but they do exist. For instance, the Hudson Institute states bluntly "There is no shortage of energy fuels" and reports the life span of various fuels as follows:

"World fossil fuels: 100 to 1,000 years supply (depending on the growth rate of demand).

U.S. fossil fuels: 50 to 100 years.

U.S. uranium (less than $100 per pound — non breeding): 80 to 300 years.

U.S. uranimum and thorium (less than $10 per pound — breeding); 80 to 1,000 years.

World Uranium (less than $10 per pound — breeding): 100 to 5,000 years."

Source: The Hudson Institute, Inc., Vol I, *The Business Environment in 1975-1985*, Paper H1-1898-RR, (New York: 1974) p. IV-10.

Shortages, according to the Hudson Institute, arise from poor planning and political friction, but "if we know that we have 100 years supply, using known technology, we can be confident about the long-term future."[1]
Public discussion has assumed an absolute shortage of energy resources despite reality. This assumption enables institutions and social engineers with a bias in planning our future to jump in and allocate objectives by arbitrary choice among fuel alternatives. Clearly, the Department of Energy and its bureaucratic predecessors have a vested interest in continuing shortage and crisis. Without a perpetual crisis, there would be no Department of Energy and no jobs for 20,000 bureaucrats.

Other institutions, including the Ford Foundation, the Committee for Economic Development and Rockefeller's Commission for Critical Choices have long favored government planning rather than the free enterprise approach to economic problems, real or imagined. The following table illustrates the way in which these government and quasi-government bodies want to allocate energy resources in the context of created shortages.

Each of these anxious energy planners, the Federal Energy Administration, Ford Foundation, Committee for Economic Development and the Commission for Critical Choices for Americans assumes an absolute energy shortage. This assumption enables their would-be planners to propose plans

(1) The Hudson Institute, Vol. I, *The Business Environment in 1975-1985*, Paper H1-1898-RR, (New York: 1974) p. IV-10.

TABLE 6-3: ESTIMATES FOR U.S. ENERGY SUPPLIES IN 1985 IN QUADS

	Oil	Natural Gas	Coal	Nuclear	Other*	Totals
Demand (1974)	33.5	22.2	13.2	—	—	73
FEA[1] Maximum Production levels (Oil $11/barrel)						
Business as Usual	31.8	24.4	24.8	7.0	4.6	91.6
Accelerated Development	42.4	29.3	47.3	8.2	8.7	135.9
Ford Foundation[2]						
High Domestic Oil and Gas	32	29	25	10	9	105
High Nuclear	32	29	23	12	9	105
Technical Fix						
Self-sufficiency	30	27	16	8	4	85
Environmental Protection[3]	29	26	14	5	4	78
CED	28.5	26.5	21.5	10	8.5	95
Commission on Critical Choices[4]	35	28	33	12	8	116

* Includes shale oil, synthetic oil and gas, hydroelectric, geothermal, solid waste, solar, etc.

(1) Federal Energy Administration, *Project Independence Report,* November, 1974.

(2) Ford Foundation Energy Policy Project, *A Time To Choose,* 1974.

for allocating use of each energy form in various proportions according to their aesthetic sensibilities. The elementary scenario "Remove Government from Energy" is never considered: the scenario is not considered because under a free market there would be no need for planners and their allocations. In sum, their estimates are dishonest and unnecessary because they are based on false assumptions.

Examine a couple of illustrations of this accusation. Richard Wilson, a professor of physics at Harvard University, presents us with a simple alternative: "Power policy, plan or panic?"[1] Wilson states his extraordinary assumptions, a vital part of the scientific approach, despite data presented in earlier chapters. Wilson's assumptions include the following:

* "Coal of good quality is nearly exhausted...Moreover coal supplies will last only 300 years,"
* "At any rate, U.S. oil supplies will last only 20 years. Foreign supplies will last 40 or 50 years but are increasingly dependent upon world politics."
* "Natural gas is as short as oil: It is harder to import and its use will probably have to be limited."

These statements have no relation to scientific fact.

We find the same "plan or perish" ultimatum in Rocks and Runyon,[2] summed up in their phrase "America faces an immediate, severe power shortage during the 1970s and 1980s.

(1) *Bulletin of the Atmoic Scientists*, May 1972.

(2) Lawrence Rocks and Richard P. Runyon, *The Energy Crisis*, (New York: Crown Publishers, Inc., 1972).

continued from previous page

(3) Committee for Economic Development, *Achieving Energy Independence*, December 1974.

(4) Commission on Critical Choices for Americans, *Energy: A Plan for Action*, 1975.

We face this disaster without any coherent national design for meaningful power procurement'' (page xii). Rocks and Runyon do present more data than Wilson but lack the understanding of economic theory for the market mechanism. Once again, government planning is the only solution considered. And even then, energy resource assumptions are far too low.

TABLE 6-4: PROBABLE LIFESPAN OF U.S. ENERGY RESOURCES ACCORDING TO ROCKS AND RUNYON

GAS	40 years at the 1970 consumption rate, less than 30 years at the present growth rate.
OIL	20 years at the 1970 consumption rate, and less than 15 years at the present growth rate.
COAL	200 to 300 years assuming synthetic oil and gas at present growth rates.
URANIUM	100 to 1,000 years after the breeder reactor.

Source: Lawrence Rocks and Richard P. Runyon, *The Energy Crisis*, (New York: Crown Publishers, Inc., 1972), p.9.

To summarize, there is a profound difference between geological data as recorded in standard research sources and the estimates printed by those who have prefabricated non-market solutions for the ''energy crisis.'' Table 6-5 compares the estimates for energy resources presented in the earlier chapters of this book with those of two pro-crisis authors.

The absurdly low estimates used by the pro-crisis schools (including the Ford Foundation and the Rockefeller Commission on Critical Choices) would have no consequence

TABLE 6-5: CONFLICTING STATEMENTS ON DOMESTIC ENERGY RESOURCES

ENERGY SOURCE	CRISIS ESTIMATES Rocks & Runyon[1]	Wilson[2]	GEOLOGICAL SOURCES[3]
Natural Gas	30-40 years	"as short as oil"	200 years
Petroleum	15-20 years	20 years	150 years
Oil from Shale	—	—	1,500 years
Coal	200-300 years	300 years	6,000 years
Atomic Power	50 years	Fusion power not before 2,000 AD fission power now	600 years, infinite with breeder reactor.

(1) Lawrence Rocks and Richard P. Runyon, *The Energy Crisis,* op. cit.

(2) Richard Wilson, *Bulletin of the Atomic Scientists,* May 1972.

(3) Chapters Two - Five.

if we adopt a market economy approach to energy production. The market system itself would balance supply shortfalls and unrealistic geological estimates simply by the mechanism of price. Unfortunately, we do not use a market system in energy; we use a politicized allocation system. The reasons are not hard to find. Notably, Rockefeller's Commission on Critical Choices found an "energy crisis" and Nelson Rockefeller was subsequently able to produce a $100 billion crash program to develop energy. It is a totally unnecessary program but obviously self-serving for the Rockefeller family interests.

We now have to turn to the results of this amateur and inefficient method of allocating who uses what fuel at what prices.

PART TWO

WHAT WE DID WITH IT....

VII

WHY IS THERE A NATURAL GAS "SHORTAGE"?

"...I became an advocate of decontrolled natural gas when I found the Federal Power Commission a few years ago turning down sales to my state of natural gas on the basis that they were 25 cents/Mcf over the controlled price. The result was we ended up buying it for $4 and $5/Mcf from Algeria or using higher priced SNG at $5 or $6/Mcf. It is absolutely insane."

CONGRESSMAN JACK KEMP,
CONGRESSIONAL RECORD, 8/3/77, pg. H8405.

Shortages cannot last long in the economic theory of a free society. If a free market system is allowed to operate without political direction, resources are allocated automatically to those uses in demand by consumers. Any temporary supply deficiency of natural gas should bring about rising prices until sufficient inducement is created to either bring forth more natural gas from suppliers or develop and market substitutes for natural gas, or to curtail demand. Similarly, any temporary surplus of natural gas will either induce a downward pressure on price until a market clearing price is established, or will cause substitutes to be abandoned and replaced by natural gas.

Where substitutes exist (and inexhaustible substitutes do exist for natural gas) the free market allocation system works to bring these into use with predictable efficiency but not perhaps always with sufficient speed to satisfy the impatient, and certainly never to take from some in order to give to others.

HOW THE SYSTEM FAILED

This system obviously isn't working. We do have shortages of natural gas. Remember 1972? True, the fall and winter of 1972 were colder than normal, the South had temperatures 15 percent or so below average, and cold weather increases demand for heating and so for natural gas, but the gas wasn't even available at the well head to fill demand. Companies on an "interruptible basis" had supplies curtailed. Ironically, some of these firms had previously converted from fuel oil to gas under environmentalist pressure. Most companies were able to find temporary alternate sources; some switched to standby fuel use, from stocks built up in anticipation of gas cut-offs; others, including utilities, were tied to a grid that was fed by coal and nuclear power plants; still others used high priced propane gas. Some firms inevitably stopped production and laid off their workers.

In Denver, the Gardner-Denver firm had no gas from early December, ran out of fuel and had no alternative but to lay off 700 workers. Even Texas, which supplies 40 percent of American natural gas, had the first major curtailments in Texas history, ranging from 50 to 100 percent. Texans, incidentally, had to pay higher prices for substitutes. As winter entered January and February 1973, supplies became critically short in many areas. In the Midwest, grain shipments were stalled because no fuel was available to move the ships. Exxon asked heating oil distributors to reduce inventories to alleviate shortages. Hundreds of schools, factories and public buildings were closed across the country, with the Midwest being the hardest hit. The winter of 1973-74 was not quite so bad with only an eight percent curtailment in interruptible gas supplies from 20 major pipelines serving the Northwestern United States between April 1973 and March 1974, and slightly greater curtailment in the following winter, 1974-75.

POLITICAL INTERVENTION THE CAUSE

Gas poverty in the midst of plentiful reserves suggests from a common sense perspective that exploration and development should be stepped up so that increments in reserves match the annual flow into the transmission system. But financing of exploration is a major undertaking and gas production firms need rate increases to cover the increased costs. Rate increases have to be approved by the Federal Power Commission, and apart from delays as the application works its way through the bureaucracy, any rate increase application is a hazardous undertaking. An increase can be challenged by users more interested in short-run low rates than long-run gas supply in a way that a market cannot be challenged. Even minor changes in a utility's operation can affect a rate increase; for example, the Orange & Rockland Utilities, Inc., moved a thermometer from the side of a building to a tower, and its application for a rate increase was confronted with a $1 million challenge on the grounds that temperature readings would now be lower than normal and the increase should not be granted. A lower temperature meant greater discounting of the utility's revenue. When a *Wall Street Journal* reporter called for further information, he was told "...there wasn't anyone around to comment because they were testifying at a rate hearing..."[1]

In brief, the major restriction on supply is not reserves (as we noted at length in Chapter Three). The restriction is governmental regulation undertaken in the name of protecting the consumer. Yet, although the cost of restriction in the form of curtailment of gas supplies might be obvious to the layman, the cost is not obvious in Washington. During every natural gas crisis some politician pipes up to demand a federal corporation to explore and drill for gas on the grounds that private enterprise is not doing the job.[2] Senator Ernest F. Hollings of South Carolina presented the emotional argument for cheap gas, wasteful usage of gas and ultimately a gas shortage as follows:

"This is the worst possible time to deregulate natural gas prices. It would further exacerbate the nation's inflation and recession problems by enlarging the oil industry's...profits...We can't allow natural gas prices to rise to the equivalent of oil prices which have been condemned as extortionate and excessive by virtually every spokesman in the western world."

So why do we have shortages in natural gas? Simply because several decades ago, ideologues with political power became impatient with the market system, imagined non-existent monopolies, decided to play God with the supply and demand of natural gas and fix maximum prices and control supply. The inevitable result has been shortages. On the other hand, if the ideologues had lavished funds on the natural gas industry, there would have been an over-supply of gas. This happened with hospital beds, after Congress voted funds for

(1) *Wall Street Journal,* July 17, 1973.

(2) See *Wall Street Journal*, December 15, 1972 for Senator Magnuson's proposal.

building hospitals, and with Ph.D.s after Federal funds were lavished on higher education in the 1960s.

As far back as 1924, the state intervened into natural gas pricing, in a way which restricted supply by substituting a political allocation mechanism for an economic mechanism. An interstate high pressure transmission developed, the Supreme Court in Missouri vs. Kansas Natural Gas Company (265 U.S. 298) placed some limitation on the power of states to intervene in gas pricing in interstate commerce; however, as a result of political accusations of monopoly on the part of pipeline companies, the Natural Gas Act passed in 1938 placed control of natural gas transportation in interstate commerce and the sale of natural gas in interstate commerce in the bureaucratic hands of the Federal Power Commission (FPC). The powers included that of establishing so-called "just and reasonable rates" for gas at the wholesale level for resale purposes. The key to subsequent shortages is that "just and reasonable prices" are man-made, uneconomic devices more than likely designed to exclude the many factors reflected in those prices formed in the impartial market place. Where regulated prices are under consumer influence, they will almost certainly be below the prices required to bring forth sufficient supplies to satisfy demand. This then becomes an indadquate price for the inducement of further development of natural gas and gas substitutes. In fact, the important function of substitutes for natural gas is nowhere considered by Congress or the Commission in natural gas pricing.

Subsequently, the Federal Power Commission moved to bring companes engaged only partially in interstate movement of gas within the regulatory fold. Congress overturned court decisions supporting the FPC, but twice legislation to return natural gas prices partly to the market place was overturned by Presidential veto. In 1945, the Supreme Court found in Colorado Interstate Company vs. Federal Power Commission (324 U.S. 581) that a natural gas company owning production and gathering facilities came within control of the FPC. Subsequently, in 1950, Congress passed the Kerr Bill to exempt certain natural gas sales from regulation. The Kerr Bill was vetoed by President Truman. Four years later in 1954, the Supreme Court placed Phillips Petroleum Company (347 U.S. 672) under the jurisdiction of the FPC, thus including all sales of gas moving in interstate commerce within FPC regulations. Once again, Congress intervened to restore a measure of the market place to natural gas, but this time the bill was vetoed by President Eisenhower in February 1956. In brief, the widening focus of natural gas regulation is a result of Supreme Court interpretation and bi-partisan Presidential vetoes. Congress has attempted at least twice to restrict the pricing powers of the FPC.

A 'FAIR AND REASONABLE' PRICE

In attempting to do the impossible, that is, to stimulate a "fair and reasonable" price for natural gas, the problem of adequate supplies has been lost in the political arguments. Some Court Justices have held that "prudent investment" is an adequate basis for gas prices. Other Justices hold that there is an artificial entity, a "fair field price" which can be ascertained and fixed by the Federal Power Commission. In all the confusion involving the Supreme Court, the Federal Power Commission, Congress and Presidential vetoes, one fundamental fact has been ignored. Any price, whatever we call it, except a market price, will create distortions in supply and demand. Almost certainly, a non-market price will yield a less-than-market clearing price and so create supply shortfalls. A market price can only be created in the market place, and any price other than a market price must create either a surplus or a deficit.

In practice, FPC determined prices paid to natural gas producers have lagged behind the consumer price index resulting in an erosion of producer earning capacity and so reducing capital formation capabilities in the gas industry. Table 7-1 shows prices paid by interstate natural gas pipelines to domestic gas producers and the consumer price index between 1964 and 1972, the decade prior to the shortages:

TABLE 7-1: CONSUMER PRICE INDEX AND INTERSTATE GAS PRODUCER'S PRICES 1964-1972

	CONSUMER PRICE INDEX[1] 1964 = 100		PRICES PAID TO GAS PRODUCERS BY INTER— STATE PIPELINES[2]	
YEAR	TOTAL	RESIDENTIAL GAS	¢/MCF	INDEX 1964 = 100
1972	134.9	123.2	20.54	123.9
1971	130.6	119.0	19.23	116.0
1970	125.2	109.3	18.11	109.2
1969	118.3	103.5	17.62	106.3
1968	112.2	101.7	17.32	104.4
1967	107.6	100.7	17.13	103.3
1966	104.6	100.9	16.87	101.8
1965	101.7	100.3	16.70	100.7
1964	100.0	100.0	16.59	100.0

While the consumer price index climbed 34.9 percent between 1964 and 1973, the residential gas component of that

(1) Department of Labor, Bureau of Labor Statistics, Consumer Price Index.

(2) Federal Power Commission.

index climbed only 23.2 percent. Prices paid to natural gas producers climbed 23.9 percent in the same period, nearly paralleling the residential gas prices. In sum, payment to producers lagged well behind the general consumer price index. Even if prices had kept in line this would not have reflected increasing scarcity of natural resources and so increasing cost of production.

THE LIMITS OF MONOPOLY

Opponents of market prices who support low prices for natural gas cite monopoly and monopoly pricing by natural gas pipeline owners as a reason for government intervention. Certainly monopolists will always act as monopolists and pipeline operators are no exception to this concept. However, a monopolist is always constrained by his demand curve, that is, how much will be bought at various prices. How is a monopolist constrained? Simply because consumers at some price will always buy alternatives, in this instance consumers will switich to alternative fuels or alternate methods of transporting gas. Pipeline gas is not a monopoly when we consider gas substitutes or natural gas not moved in a pipeline or the production of gas at the field level. Bottled propane, locally manufactured gas and imported gas are competitive alternatives if domestic pipeline operators get too far out of line in prices. Paul W. MacAvoy states bluntly that the assertion of monopoly "has always no basis in fact" at all.[1]

(1) Paul W. MacAvoy, *Price Formation in Natural Gas Fields*, (New Haven: Yale, 1962) pp. 5 and 265.

"As has been suggested by this long analysis of price formation in the 1950s, gas markets were diverse in structure and behavior, and were generally competitive or were changing from monopoly toward competition. It would seem possible that this could result in price level changes and in a revolution of pricing patterns in contrast with those associated with monopoly. Markets with such characteristics need not be regulated by the Federal Power Commission to prevent monopoly pricing. Nor, at present, is it easy to justify regulation on other grounds, given the undesirable effects on excluded consumers and the rate of discovery and sale of reserves." (p. 265).

Even if the monopoly pricing argument had actual justification and if planners were able to simulate market prices in the regulatory process, the idea of supply by plan has other less noticeable, but equally harsh defencts, notably the results of imperfect information. Limitations of planning can readily be seen in the coal gasification process. The 1952 Paley Report was apparently unaware of coal gasification; the gasification process is not described and the Paley Report contains the comment:

"Nor is there yet developed for gas, as for liquid fuels, a substitute derived economically from this country's abundant reserves of solid fuels...."[1]

In fact coal gasification was known long before 1952 and used in Germany and Belgium and was called the Lurgi process. Obviously, if market clearing prices determined by the market place had been used in natural gas, private initiative would have explored the technology and economics of coal gasification for the United States. The process was not

(1) The President' Materials Policy Commission, *Resources for Freedom: Foundations for Growth and Security*, (Washington, D.C.: 1952), 1, 111-114.

explored in the United States because the price of natural gas was artificially depressed by the FPC, thus encouraging wasteful use of gas. For example, in production of carbon black, the Paley reported noted that "because of low prices," the carbon black industry "makes notoriously inefficient use of the heat and carbon content of natural gas which already is under partial state regulation."[1]

Only twenty years later did the 1973 Klaff Committee[2] in its final report (page 4B-8) take implicit note of the role of pricing in determining gas supply. Natural gas is listed with reserves of only 279 trillion cubic feet at 1971 prices, but in absolute terms, ignoring price, with "moderate" identified reserves and "large" hypothetical reserves. In sum, the Klaff Report recognizes that shortages of natural gas are non-economic creations generated by political manipulation of prices.

WHO WANTS REGULATION?

A study of this relationship between ideology and urge to regulate natural gas prices was made by Edward J. Mitchell of the University of Michigan. Mitchell found a distinct correlation between ADA (Americans for Democratic Action) rating and votes on regulation of natural gas. ADA is, of course, interventionist in the Keynesian tradition and vocally in favor of gas regulation, supposedly in the interest of the consumer. Mitchell examined the Congressional vote on an amendment to natural gas prices. The relationship between

(1) *Ibid.* p. 114.

(2) National Commission on Materials Policy, *Material Needs and the Environment Today and Tomorrow,* (U.S. Congress, Washington, D.C., 1973).

Congressional votes on deregulation and ADA rating was as follows:

TABLE 7-2: COMPARISON OF ADA RATING AND DEREGULATION VOTING RECORDS

ADA RATING	VOTES AGAINST DEREGULATION	VOTES FOR DEREGULATION
95-100%	26	0
85-90	31	0
75-80	41	4
65-70	33	2
55-60	30	3
45-50	19	10
35-40	8	18
25-30	7	20
15-20	1	37
5-10	3	69
0	0	37

Source: *Wall Street Journal,* July 7, 1977.

According to Mitchell, if a Congressman's rating with ADA is 45 percent or higher, he will vote against deregulation, presumably under the mistaken belief he is helping his constituent consumers. Thus, the New Jersey delegation voted 14:1 against deregulation — and New Jersey has been one of the worst hit states in the natural gas shortage. At almost the same time Congress was voting against deregulation, President Portillo of Mexico suspended a Mexican offer to sell natural gas to the United States. Why? Because the U.S. refused the Mexican asking price, $1 higher per MCF than the depressed U.S. domestic price.[1] Interventionism into the marketplace is

(1) *Wall Street Journal*, December 23, 1977.

even more widespread than Mitchell suggests. For example, Edward Teller, usually thought of as a conservative, wrote a report for the Rockefeller Committee on Critical Choices for Americans and decided that the target for an *in situ* gasification from coal program should be established at two quads of natural gas:

"We recommend the production of two quads from *in situ* gasification of coal. Just as the vigorous production of oil may result in the depletion of our oil reserves justifying the accelerated development of shale oil, the danger of depleting our gas reserves through sharply stepped-up production of natural gas should be compensated by the development of *in situ* gasification of coal. By this technique we could exploit deep, thick coil veins which are hardly accessible or are inaccessible with present methods." [1]

Obviously there is no way Teller can determine precisely the amount of natural gas required several decades in the future. The figure of two quads may be either too high or too low. This inability was recognized in the Foster Associates, Inc., study, commissioned by the American Petroleum Institute and which obviously would have made the claim if it could legitimately be made. On the contrary, the Foster report states:

"The supply of natural gas is responsive to price, but no technique has been developed for making reliable estimates of the amount of new natural gas supplies which would be elicited under alternative pricing policies.[2]

The slower reaction time of government planners, compared to market system initiatives, can also be identified in

(1) Edward Teller, *Energy: A Plan for Action,* (Commission on Critical Choices for Americans, New York: 1975), p. 24.

(2) *Gas Supply Review*, September 15, 1973, American Gas Association, p. 81.

coal gasification and is a reason to reject the arbitrary Teller-Rockefeller targets. Planners first need shortages to appear as the signal for planning action. Private enterprise looks ahead of shortages as anticipation of the future is the essense of entrepreneurial success. For instance, the Bureau of Natural Gas only noted the possibilities of coal gasification in 1971,[1] and a gasification program was included only in the 1971 President's Clean Energy Message of June 14, 1971:

"The current natural gas supply shortage has served to focus attention on the character and magnitude of the United States fossil fuel resource base and the extent to which gas produced from coal gasification could become a significant portion in our energy economy." [2]

The message adds, "the desirability of converting a portion of our coal resources into a supplemental source of gaseous fuel is obvious." Translated from ponderous bureaucratese this means that the Bureau of Natural Gas and the President's energy advisers consider that the natural gas shortage is a signal to go ahead with coal gasification. It takes emergence of the problem to bring action, and even then the action is likely to create further problems.

GAS GEOLOGY STOPS AT THE BORDER

A dramatic reflection of the results of political action can be seen in gas and petroleum drilling in Lake Erie; the

(1) Federal Power Commission, *National Gas Supply and Demand 1971-1990,* (Bureau of Natural Gas, Washington, D.C., 1972).

(2) *Ibid.* p. 84.

U.S.-Canadian international border runs down the center of the lake. In Canada neither the federal nor provincial governments, although not completely free of bureaucratic interference in mining, have restricted drilling in the lake and in 1975, one-half of Canadian gas came from Lake Erie.

Unfortunately, Lake Erie has been leased out to gas and oil drillers only on the Canadian side of the border. What happened in the U.S. part of the lake? No gas is produced from the U.S. sector. Did the gas formations cease at the international border? On the contrary, the geological formations are favorable but back in 1968, both New York and Ohio halted offshore drilling because of alleged water pollution. (Presumably Canadians are not sensitive to this kind of water pollution). Regulatory hearings commenced, and it was not until 1977 that New York, which imports 95 percent of its natural gas, allowed drilling. Even now, government restrictions imposed are so onerous as to make it uncertain whether gas producers will be interested in drilling the New York sector of the lake.

To put the shortage of natural gas briefly: we have shortages and will have more shortages because natural gas prices are political events remote from the market place. As these political events are under consumer pressure to be lower than market clearing prices, we know that gas will be wasted in uneconomic uses, that gas substitutes will not be developed and that gas exploration and development will not take place. What about the likelihood of monopoly pricing? Over the long term this is impossible. Too many substitutes exist for natural gas and production is too diverse and scattered to yield a tight monopoly. The basic causes of the gas shortage have nothing to do with energy: they stem from a philosophy of something for nothing, where the consumer wants gas at less than the cost of production and a political structure where politicians are willing to pander to the crowd to get reelected.

CHART 7-1: LICENSES AND LEASES IN THE CANADIAN PART OF LAKE ERIE

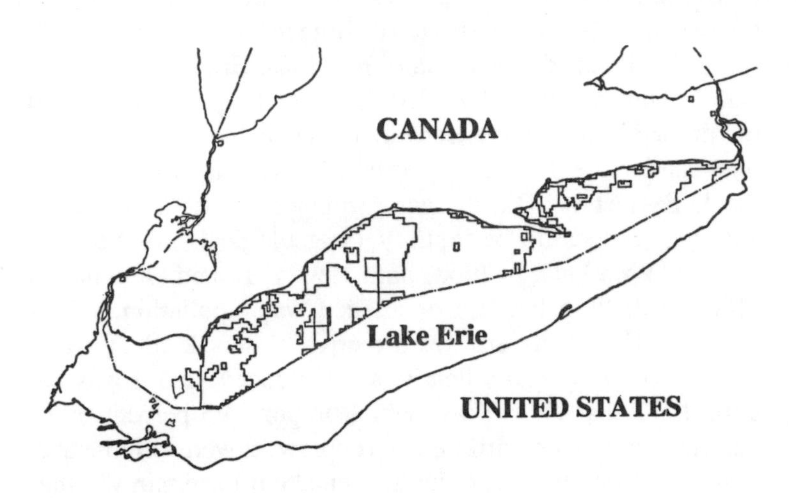

Source: Ontario Ministry of Natural Sources, *Ontario Mineral Review: 1975*, (Toronto: 1976), p. 35.

Finally, it is not beyond possibility that political power groups, such as the Commission for Critical Choices for Americans (in the Teller-Rockefeller report) welcome shortages as a vehicle to translate their program for New World Order into reality.

The solution to the natural gas dilemma? Prompt, complete deregulation of all natural gas production and distribution.

Chapter VIII

VIII

IDEOLOGY IN THE MARKET PLACE

"When people (Congressmen and their constituents) do not understand an issue they fall back on ideology. By ideology I mean that general body of beliefs you consult when you have to make a judgment on something you know little or nothing about."

EDWARD J. MITCHELL, PROFESSOR OF BUSINESS ECONOMICS, *WALL STREET JOURNAL*, JULY 7, 1977.

Politicization of decision making in energy is at the root of the so-called energy crisis. Undoubtedly, important contributory factors include an abysmal lack of economic understanding among policy makers, would-be planners and those who see energy as a vehicle for promotion of self-interested boondoggles. The problems of politics and ignorance are compounded by the authoritarian approach to economic teaching in major universities during the past several decades, a period in which those disputing Keynesian wisdom have been purged as students and as teachers. Pure ignorance of basic classical economics in the political field is a potent prescription for disaster. Most of our "experts" have received their economic education from Keynesians and neo-Keynesians, who have held a monopoly in college teaching during the years since World War II. It is doubtful if any Ph.D. in macroeconomics has been awarded in the last two decades to a graduate student of any other theoretical persuasion. It is now more obvious that Keynesianism is plagued by economic fallacies, and we can count the so-called energy crisis as one of these.

At one perspective, energy policymakers assert that energy resources are scarce in some absolute way; therefore we must conserve. From another perspective, these same policymakers act as if resource scarcity did not exist by pricing below market clearing price, by inefficient government regulations, and by various giveaway schemes. This inconsistency betrays a lack of fundamental economic knowledge and is fostered by the pragmatic rule book of the political world.

Standing by are those numerous and anxious special interests, ready to profit at public expense, and the multinational oil companies are among the worst of these wealthy beggars.

ECONOMICS VS. POLITICS

Classical economic theory is still the most useful explanation of economic events. In classical economics theory there is no such event as a shortage. A shortage cannot exist in a freely functioning market place. A shortage results only from political intervention into the market place for ideological reasons, and such interventions arise just as readily from anti-free enterprise big business (Atlantic Richfield is one such multi-national) as from smaller groups. Ideology is the only reason shortages arise.

Consequently, political intervention into the energy market place since the early 1920s has been the major factor in creating the present crisis. Moreover, this political intervention came not only from politicians wooing voters but from the energy industry which is fearful that the impartial market place is too cold an environment and the political route, which can be swayed by lobbies and dollars, is an easier way to generate and guarantee profits. The fault is just as much with a politically-oriented industry as with vote chasing politicians and self-serving social engineers.

Natural gas is an example where an industry has been pilloried for ideological reasons. The Natural Gas Act of 1938, confirmed by the Supreme Court's 1954 Phillips Petroleum decision, placed natural gas pricing in the hands of the Federal Power Commission bureaucrats. With what result? If the bureaucrats could establish market prices for gas, then bureaucrats would be superfluous. A bureaucratic price is either above or below a market clearing price. Consequently, the supply generated through the regulated prices will be either greater than, or less than, the demand. Under political pressure, natural gas prices have always been set below the market clearing price and so created shortages. Sufficient supply was not produced to meet demand at that price. What is the politician's answer to a politically induced shortage? The usual answer is rationing. However, rationing never solves a shortage, it only perpetuates a shortage.

Two economic distortions flow from pricing at less than market clearing prices: (a) use of the energy source is encouraged because of the lower price, and (b) supply is not forthcoming at the low price. One distortion is to facilitate use of the scarce energy source and highlights the pitiful lack of economic knowledge among environmentalists and politicians. These groups support low prices to protect the economically deprived, but the lower price leads directly to waste because more of anything is demanded at lower prices. Concerning waste of resources, books have been written to demonstrate and propagandize non-economic methods to prevent waste without noting the fundamental fact that the problem of waste can not even exist if the non-economic approach is abandoned. Thus, energy waste is ideologically created waste. A group of authors from major universities compiled *Resources and Decisions,*[1] a catalog of energy waste and their recommendations for eliminating it. Extensive calculations are made for recycling and the effects of recycling, but not one word to the effect that recycling will come about automatically if it is economically feasible. Instead of returning to the market place, where substitutions at the margin of higher prices automatically bring about rationing of scarce resources, the Leonardi scholars adopted confused and complex answers guaranteed to perpetuate the energy problem.

"subsidies could be extended to the secondary materials industry. These might be tax credits to manufacturers who increased use of recycled materials; fast tax write-offs for expensive scrap processing machinery, such as automobile shredders; low cost loans or guaranteed capital to secondary materials industries; or price supports for secondary materials through federal purchase and stockpiling." 2

UPSIDEDOWN ECONOMICS OF HEATING OIL

Practical examples of such economic illiteracy can also be found in government decisions and actions. In the winter of 1972-73 there was a shortage of heating oil in the eastern

(1) The Leonardi Scholars, Duxbury Press.

(2) *Ibid.*, p. 111.

regions of the United States. As we have already pointed out, in a free market system, there can be no such event as a shortage. A shortage is an indicator. It indicates that prices are not sufficient to induce a supply to equate demand. The solution would be to raise prices until it induces a supply of heating oil to eliminate the shortage. But in February 1973, right in the midst of the energy crisis and a major shortage of fuel oil in the United States, the White House announced a program. Not a program to increase prices and so expand supply, but a program to reduce prices of heating oil. The result is obvious to any first year economics student. A reduction in price will aggravate the shortages. And this is precisely what happened.

Moreover, the Administration didn't just request a rollback in prices. It got tough about it. Treasury Secretary Schultz threatened to discipline violators with the "club in the closet."[1] Subsequently, in early January 1973 there was a critical shortage of kerosene fuel for jet aircraft at New York airports: both Trans World Airlines and American Airlines were forced to make unscheduled fuel stops outside New York to pick up kerosene unavailable in New York. Why was no jet fuel available at New York? The immediate reason was that facilities normally used for storing jet kerosene were filled with heating oil. Why were the tanks filled with heating oil? Because the Federal Government had requested storage of heating oil to avoid a heating oil crisis. But the weather was warmer than the omnipotent Federal Government anticipated and the heating oil in the storage tanks hadn't been removed to make way for the jet kerosene. Now private industry under non-interventionist free market economics has operated for decades without making that stupid kind of error.

(1) *Wall Street Journal,* February 8, 1973.

BIG BUSINESS COLLECTIVISTS

Amid the clamor for redistribution from the haves to the have nots for free handouts, the subsidies and schemes to ban atomic energy and breeder reactors, substitution of natural gas and oil burning for coal (and then switching back to coal) one voice should be heard in favor of rational solutions: that of American business.

Certainly, the energy industries — and Mobil is one prominent voice — have placed themselves on record in favor of a market system and are very critical of the environmentalist, no growth, anti-technology syndrome. However, business is inconsistent in its approach to the market system. Indeed, it has almost neutralized its own voice. For instance, back in 1973 the energy industry issued a Joint Energy Policy statement.[1] Regrettably, instead of a reasoned approach emphasizing the abundance of undeveloped domestic energy resources, the report emphasized the scare cliche "critical energy program," formulates government planned solutions such as "energy problems must be placed high on the list of our national priorities," and so on. Within the report, the industry group acknowledges that there is "no shortage of resources" and does suggest that government interference is the prime cause of any crisis including "stringent environmental standards" and "government efforts to superimpose its direction as a substitute for market forces."

(1) *Toward Responsible Energy Policies,* published jointly by American Gas Association, American Petroleum Institute, Atomic Industrial Forum, Inc., Edison Electric Institute and the National Coal Association (March 12, 1973).

However, instead of urging a return to market forces as a solution, the joint report merely suggests "steamlining" of government regulations to the extent that government policies should recognize "the benefits of the free enterprise system." Moreover, the industry report supports government intervention to its own benefit. It wants government "incentives," "national land use policies" and import policies designed to "promote" domestic development. Finally, the industry urges "long range governmental commitment to research and development in the energy field." In sum, the major U.S. energy industry elements propose policies no different from welfare recipients, except that the welfare is to be directed to the energy industry and not to impoverished consumers.

American business is shortsighted. It is unable to see its true long-term interests in a wider perspective. Businessmen believe that they can grab government grants and manipulate government policies to their own objectives and yet avoid being called socialists. The impact of the energy appeal will be negative, and correctly labelled as self-serving.

The American Gas Association booklet *Elements of an Effective Energy Program* does encourage abandonment of government regulations that inhibit production but, a few paragraphs later, calls for government assistance for energy industrial research, which even today almost equals industry expenditure:

Out of this collectivist jungle only one clarion call has emerged — from Mobil Oil in a remarkable set of articles, in March and April 1977, which called for abandonment of a political approach to energy and use of the market system. Only in a few paragraphs did Mobil Oil depart from its clear-headed challenge.[1]

(1) *A National Energy Policy*, Mobil Oil Company, 1977.

TABLE 8-1: EXPENDITURES OF ENERGY R&D IN FISCAL YEAR 1974 (in millions of dollars)

	GOVERNMENT	INDUSTRY
Coal	$105.8	$10
Oil and gas	9.1	710
Nuclear	566.8	300
Solar	12.2	negligible
Magnetohydrodynamics	5.1	negligible
Pollution control technology	46.0	negligible
Miscellaneous	9.5	negligible
TOTAL	$745.5	$1,020

Yet in looking at Washington lobbying activities, we find even Mobil plays the double standard game with no fewer than five full-time Washington lobbyists. To keep the flow of federal largesse pouring into the energy industry, and effective and persuasive Washington lobby is maintained. The oil industry is a typical example. In 30 states the oil industry has trade associations. These associations are backed up by regional organizations with a prime purpose of monitoring oil industry legislation; for instance the Western Oil & Gas Association covers the Pacific Coast states. In addition, there is the powerful American Petroleum Institute with about two hundred employees. This active lobbying organization, highly sensitive to the demands of the industry, is backed up by individual oil company lobbyists:

> None registered for Occidental, Cities Service, Ashland, Kerr-McGee and Marathon. One each registered for: Continental, Stand Ohio, Tenneco, Texaco, Getty, Superior, American Petroleum Institute, American Gas Association, Gas Supply Committee.

Moreover, petroleum interests are represented within the executive branch of government itself by former oil in-

TABLE 8-2: OIL COMPANY LOBBYISTS REGISTERED AS OF MARCH 1974

Exxon	7
Shell	7
Atlantic Richfield	7
Mobil	5
Gulf	4
Standard Indiana	5
Phillips	3
Union	3
Sun	4

Source: Norman Medwin, *The Energy Cartel*, (New York: Marine Engineers Beneficial Association, 1975), p. 53.

dustry employees working temporarily in Washington through what is known as the "revolving door." In 1974, for instance, no fewer than 102 former oin industry employees worked at the Federal Energy Office, including ten in supergrade positions GS 16-18 and another 59 in grades GS 13-15. There is an incestuous relationship between the energy industry and the executive branch of government.

Resulting from this "revolving door," research contracts are concentrated among those corporations most politically active. For example, the American Gas Association's methane gas from kelp farm is run by General Electric. Also, a Federal $7.7 million contract for research on reducing "flow indeed" vibration in nuclear reactors was won by General Electric. Thus, G.E. keeps in the forefront of two important technical areas with taxpayers' money. In fiscal year 1978-79, $3 billion will go into an incredible array of energy solutions. Some will prove commercially successful, some will fizzle, but in the absence of a free market the choice will be inefficient. There are subsidies for making electricity from the sun, from chemical fuel cells, from hot gas in a magnetic field, for tapping energy from ocean currents and giant windmills and underground streams. *Forbes* magazine estimates that the Federal government now pays two-thirds of the cost of pilot projects and one-half the cost of energy demonstration plants. A key statistic is that three quarters of the Department of Energy budget is assigned to the Energy Research and Development Administration (ERDA) for research projects that should, in a free market, be funded by private industry. More than $2 billion is currently planned for nuclear research, and $1.4 billion for coal gasification and liquification. Obviously, the energy bureaucrat's most enthusiastic support is big business. The efforts of the zero growth activists, the anti-nuclear protestors, the conservationists, all look insignificant against these big business bankrolls. Big business likes to talk private enterprise and market economy. Individual businessmen get livid if labelled socialist, bit in fact the energy industry is one of the greatest recipients and one of the most active promoters of federal largesse.

THE ECOLOGISTS ATTACK BIG BUSINESS

In many ways, the environmentalist attack on the oil companies is just as reprehensible. For instance, Robert Engler in *The Politics of Oil,*[1] states that "not one major new refinery was constructed on the East Coast in the fifteen years prior to 1973." This is true. But the relevant question is why were no new refineries built.

There has been no new construction because new refineries require coastal sites for the most part, and environmental opposition to coastal sites is persistent and unreasoning. In Delaware, the state legislature, under environmentalist

(1) Robert Engler, *The Politics of Oil,* (New York: The MacMillan Company, 1961).

pressure, has refused to allow a refinery to be built anywhere along the Delaware coastline. Hazy Federal air quality standards and Federal regulatory authority indecision, encouraged by the ecologists, halted refinery construction in other areas. You don't build a refinery until the specification of gasoline to be produced is known. Without air quality standards, you can't determine the gasoline specification. So while the environmentalists complain of shortages, they are themselves primarily responsible for the shortages. This is not said to whitewash the oil companies. But the oil industry should not be blamed for events for which it is not responsible. Refineries will be built when it is profitable to build refineries.

This compounding of a politically created energy shortage is a highly vocal and well-funded effort, quasi-political in nature, using protection of the environment as a device to reduce consumer access to energy and reduce living standards. These efforts are absurd to the extent they place the well-being of birds and fishes above the welfare of human beings.

Ford Foundation financed studies of the effects of oil spills on marine environment and these illustrate the extent to which occasional and avoidable problems can be escalated into a device to reduce overall living standards. According to one Ford Foundation report [1] marine life can be killed in the following ways:

"(1) Coating and asphyxiation (example: barnacles and other intertidal life);
(2) Poisoning through direct contact or ingestion (examples: ingestion of oil by preening birds, contact poisoning of vascular plants);
(3) Exposure to water-soluble toxic petroleum components (example: Subtidal fishes and invertebrates at Midway Island, the Tampico Maru spill, and West Falmouth);

(1) Donald F. Boesch, et al., *Oil Spills and the Marine Environment,* (Cambridge, Mass.: Ballinger Publisher Company).

(4) Destruction of more sensitive juvenile forms (example: fish eggs and larvae); and
(5) Disruption of body insulation of warm blooded animals (example: diving birds)."

This Boesch report, which is accurate in itself, was financed by the Ford Foundation. What is the solution offered? Not to avoid spills creating the conditions cited but an emphasis on government planning far beyond solutions to "coating," "poisoning," "exposure" and "disruption of body insulation."

The proposed solutions include:

"The ecological effects of oil pollution on the marine environment will be an important consideration in energy policy decisions in the future. Public pressures and legal mandates, such as the National Environmental Policy Act and the Federal Water Pollution Control Act, will insure this. Changes in policies governing oil imports will affect the possibilities of accidental spills. International agreements concerning intentional shipping discharges will be formulated. Decisions will be made on where to allow offshore oil exploration and production, and on the types of pollution prevention technology required in these production fields. Superports will be planned, as will coastal refineries."[1]

The lengths to which social planners in the environmentalist movement will go is illustrated by the proposed Mackenzie Valley pipeline in Canada. This is currently held up because Federal Commissioner Tim Berger claims the pipeline may adversely affect a herd of caribou — all 5,000 of them. What are the rights of millions of human beings, potential consumers of the oil and gas brought to the market in these pipelines compared to the rights of 5,000 caribou?

(1) *Ibid.*, p. 45.

THE ZERO RISK MANIA

Perhaps even more emotional than the environmentalists are the zero risk advocates: those who would stop all economic growth, all increases in standards of living and condemn the world to a static Orwellian 1984.

We live in a world of risk. Human survival depends on our willingness and skill in analyzing and dealing with risk situations. There cannot be a totally risk free environment; we can only reduce the risks in any given situation. The attack on atomic reactors in particular is based on a biased interpretation of risks involved in mining or obtaining the fuel needed to produce energy and the risks involved in conversion of that fuel to electricity or motive power: that is the starting point. Table 8-3 compares statistics for mining operations, fatalities and non-fatal injuries in two energy associated mining operations:

TABLE 8-3: U.S. FATAL INJURIES IN ENERGY RELATED MINING

	1960	1970
Coal Mines	325	260
Petroleum & natural gas	82	134

Source: *Statistical Abstract,* 1975, p. 681.

Injuries in Table 8-4 for energy related mining follows the same pattern:

Reading from the tables, we note that there were 260 deaths and 11,552 injuries in mining coal in 1970. In 1976 alone, 63 people were killed and 366 injured in pipeline

TABLE 8-4: NON-FATAL INJURIES IN ENERGY RELATED MINING

	1960	1970
Coal Mines	11,902	11,552
Petroleum & natural gas	9,110	9,989

Source: *Statistical Abstract*, 1975, p. 681.

accidents. When we compare these risks with those involving nuclear plants, we find the latter have an extraordinary safety record, in distinct contrast to the mob assault on nuclear plants. The Rasmussen Report [1] calculated annual fatalities and injuries to be expected for 15 million people living within 20 miles of U.S. reactor sites from various accidents, including nuclear reactor accidents. Nuclear accidents are estimates only because in two decades of nuclear plant operation there has never been a fatal accident.

The Rasmussen study compared the risk of accident in the 100 nuclear power plants expected to be in operation by 1980 with natural events where we already have known and measured risks. Although more than 60 nuclear plants are in operation, and have been for two decades, one cannot measure the actual risk of nuclear accident because there have been no nuclear accidents resulting in a significant release of radiation. This is in distinct contrast to the noisy fringe element, duly reported in the mass media, suggesting many thousands are in imminent danger of perishing from improperly functioning nuclear plants. The study concluded that non-nuclear events,

(1) The most thorough examination of reactor safety conducted by Norman C. Rasmussen of M.I.T. in a $4 million study sponsored by the Nuclear Regulatory Commission.

TABLE 8-5: ANNUAL FATALITIES AND INJURIES EXPECTED AMONG THE 15 MILLION PEOPLE LIVING WITHIN 20 MILES OF U.S REACTOR SITES FROM VARIOUS CAUSES

ACCIDENT TYPE	FATALITIES	INJURIES
Automobile	4,200	375,000
Falls	1,500	75,000
Fire	560	22,000
Electrocution	90	—
Lightning	8	—
Reactors (100 plants)	0.3	6

Source: *Rasmussen Report,* p. 19.

including air crashes, hurricanes, tornadoes, dam failures, chlorine releases, explosions, meteors and earthquakes are about 10,000 times more likely to yield fatalities than nuclear plant accidents. Furthermore, property damage is 100 times more likely to be caused by non-nuclear events, either man-made or natural, than by nuclear accidents. The risk of fatality from various causes contrasted to the risk from nuclear reactor accidents is contained in Table 8-6.

ARE NUCLEAR REACTORS SAFE?

No, nuclear plants are not safe in the sense that accidents can never occur. Even with a 1 in 300,000,000 chance of fatality there is still a certainty that at some point a fatality will occur. But neither are airplanes or automobiles completely safe. The literature on nuclear safety, especially the literature

TABLE 8-6: RISK OF FATALITY BY VARIOUS CAUSES

ACCIDENT TYPE	TOTAL INDIVIDUAL CHANCE	
	NUMBER	PER YEAR
Motor Vehicle	55,791	1 in 4,000
Falls	17,827	1 in 10,000
Fires and Hot Substances	7,451	1 in 25,000
Drowning	6,181	1 in 30,000
Firearms	2;309	1 in 100,000
Air Travel	1,778	1 in 100,000
Falling Objects	1,271	1 in 160,000
Electrocution	1,148	1 in 160,000
Lightning	160	1 in 2,000,000
Tornadoes	91	1 in 2,500,000
Hurricanes	93	1 in 2,500,000
Nuclear Reactor Accidents (100 plants)	0	1 in 300,000,000

Conversely, Petr Beckmann in *The Health Hazards of Not Going Nuclear,* [1] states:

on the possibility of nuclear reactor accidents, is polarized between a "catastrophic" school and a "nuclear energy is safest" sthool of thought.

Denis Hayes of the Worldwatch Institute in *Nuclear Power: The Fifth Horseman* [2] makes the statement:

"During the brief period since the arrival of commercial nuclear power, there have been many reactor accidents — some of them potentially catastrophic."

(1) Denis Hayes, *Nuclear Power: The Fifth Horseman,* Worldwatch Paper 6, (Washington, D.C.: Worldwatch Institute, 1976), p. 32.

(2) Petr Beckmann, *The Health Hazards of Not Going Nuclear,* (Colorado: The Golem Press, 1976), p. 81.

"For hundreds of reactor-years, there has not been a single reactor related fatality in the generation of commercial power anywhere in the United States."

The official statistics make Beckmann right and Hayes wrong. The distance of the environmental nihilists from reality is occasionally breathtaking. Barry Commoner, who finds his way to anti-nuclear demonstrations across the country to urge on the faithful, is, according to his own statements, wasting his time because, he says, the whole nuclear program is an exercise going nowhere:

"What emerges from these considerations is the liklihood that the entire nuclear program is headed for extinction. It will leave us with a monument which people will need to care for with vigilance if not affection for thousands of years — stores of intensely radioactive wastes and the powerless, radioactive hulks of the reactors that produced them.

Commoner's speeches at protest sit-ins are more inflammatory. At the Diablo Canyon nuclear plant in Morro Bay, California, which was ready to produce electricity and was being held up only by regulatory paper work, Commoner proclaimed, "This day symbolizes the destructive death-dealing power of modern technology." [1] Production of electricity from industrial grade plutonium is confused with death-dealing weapons grade plutonium in Commoner's mind. As Commoner pointed out, a few paragraphs later in his harangue, the Diablo Canyon facilities do no more than boil water. But that isn't good enough for Commoner, water must be boiled according to the Commoner formula, which is:

"You've got to match the energy to the task. You don't need radio-

(1) *Sun-Bulletin,* Morro Bay, August 11, 1977.

activity to boil water. You can do it out here with a couple of mirrors."[1]

In the final analysis, one fact stands out. In 1976 some 63 people were killed in pipeline accidents and four times that number in coal mine accidents. Yet there are no demonstrations about safety in pipelines or coal mines. There have been no civilians killed in nuclear reactors last year or any year — yet this is where the noisy demonstrations are focussed. A logical question might be — who is promoting the demonstrations and what is the monetary gain from a ban on nuclear energy?

The most reliable answer to this question is found by asking another question — who would profit from denying the American public plentiful, cheap energy.

The obvious major losers would be major oil companies who own large segments of the coal industry, but own only one quarter of U.S. uranium deposits and scarcely any overseas uranium deposits. There is far more profit in oil and coal than in cheap, clean nuclear power.

When we probe the question of profit a little further, we find that studies purporting to decide the energy future of America were undertaken and publicised by foundations controlled by the oil barons. For example, the Critical Choices for Americans energy study was funded and presided-over by Nelson Rockefeller. Rockefeller family interests control Exxon and other major oil companies. The name Rockefeller is almost synonomous with oil. While proposals from this self-interested source would appear to be *prima facie* suspect, no one appears to have challenged these plans on the grounds of naked self-interest.

(1) *Ibid.*

HOW TO MAKE 'THE' BOMB

One realistic concern expressed by counter-nuclear forces does carry weight — a crude, but possibly effective atomic bomb can be manufactured from plutonium stolen from nuclear plants. The difficulties have been underestimated; terrorists can more easily obtain other effective weapons, but a private nuclear bomb could be assembled. In fact, groups of science students in both the United States and Great Britain have written out the theoretical requirements for manufacture of a crude atomic weapon from information gained in publically available literature. These exercises are based on the supposition that a terrorist group can steal sufficient plutonium to make a bomb. In practice, weapons grade plutonium (plutonium-239) is well guarded. More likely, a terrorist group would attempt to use industrial plutonium, contaminated with plutonium-240, and with this grade the explosion can be extremely weak, a risk not acceptable in weapons systems but quite acceptable for a terrorist. In other words, there is a risk that a crude, very weak atomic bomb can be manufactured from industrial grade plutonium. This would be a weapon about the size of a tea chest, weighing some 1,000 pounds, and capable — if exploded — of killing everyone within a half-mile radius, with major blast damage up to one mile. The cost would be perhaps $30,000 to $50,000 if plutonium, explosives and materials were available, plus a half-dozen skilled technicians with a knowledge of nuclear material, explosives and nuclear chemistry. A group of university students can write up the theoretical steps for manufacture of a bomb but this is much removed from actual manufacture. Manufacture requires a knowledge of industrial processes unlikely to be found in most terrorist organizations. However, what appears to have escaped all discussion of do-it-yourself-bombs — even in the most radical of anti-nuclear discussions — is that terrorist gangs have been, and presumably still are, controlled by some major government. The Soviet Union supplies Middle East terrorist organizations directly and through Libya. The United States, through the CIA, has used terrorist gangs for its own purposes. What is more likely is that a state already in the nuclear club may funnel materials and skills to a front organization. To suggest that no responsible country would undertake such action is to by-pass an important fact: most major governments have demonstrated willingness and ability to undertake major death-dealing missions. The risk is not from a group of anti-nuclear protesters. The risk is from an established major power government setting up a terrorist organization as a weapon.

WHY NUCLEAR PLANTS ARE SAFE

The thrust of the anti-nuclear campaign is that nuclear plants are unsafe and nuclear reactors, by implication, either explode killing neighboring inhabitants or are likely to spew death-dealing radiation across the surrounding countryside.

The truth is, however, that nuclear plants cannot, under any circumstances, explode. And they are most unlikely to release any radiation either.

The complete reasons why nuclear plants cannot explode are set down in a brief and readable booklet by Petr Beckmann, a professor of engineering, entitled *The Health Hazards of Not Going Nuclear*, (Golem Press, Boulder, 1976). Consider these points described by Beckmann:

—The chain reaction in a power reactor is quite different than that in a nuclear bomb even a runaway reactor cannot explode.

—Nuclear grade uranium is more than 90% U_{235} but a power reactor uses low grade 3% U_{235}— and there is no way an explosion can be induced in uranium of this low grade.

Compare these facts to a book entitled *We Almost Lost Detroit* [1]. The book was based on a scare fantasy, that is, because the coolant system in a Detroit reactor failed and some fuel rods melted, Detroit was in imminent danger of being blown off the face of the earth or flooded with contamination. In fact, there was never any possibility of explosion and the radiation was contained normally in the containment building. In brief, the essence of the anti-nuclear campaign is a dishonest substitution of mythology for scientific fact.

THE FREE LUNCH CROWD

Energy is today remarkably cheap compared to his-inefficient production and use of energy. By a series of open and hidden subsidies, we have developed a way of life based on low cost energy. Now that the cost of these subsidies threatens to drown us, the prospect of losing cheap energy is alarming both politicians dependent on votes and those citizens who prefer welfare.

Energy is today remarkabley cheap compared to historical energy prices in the United States and energy prices elsewhere in the world. To buy a tank of gasoline today requires less than two hours work for the average worker in manufacturing compared to three hours in 1945. Since 1947, wages have risen 311 percent and gasoline prices only 176 percent. In one sense, the much-abused depletion allowance is a subsidy for gasoline prices, while it also holds back develop-

(1) John G. Fuller, Readers Digest Press, 1975.

ment of domestic oil and pipeline projects — and is furthermore a cost not reflected in present prices. In brief, we have a vast network of gimmicks threatening to partially collaps the economy. The railroads, an efficient system for freight, have been engineered into decline by an Interstate Commerce Commission which refused to allow railroads to charge efficiency prices or to earn enough to survive. Congress completed the job of eliminating railroads by subsidizing other forms of transportation using fossil fuels. An overview of energy crisis literature highlights an extraordinary fact: there is a notable lack of economic analysis prior to political action.

Arguments are couched in terms of selective information coupled with derived opinion. The consequences of scarcity are rarely considered. For example, *Energy Strategy: Not What But How*, published by a group of M.I.T. associated technicians, has no concept of economic analysis at all. The book predicts crisis merely from straight line projections of contemporary energy usage. It is a static, rather than a dynamic, model.

One cannot rationally debate energy use outside the fundamental observation of scarcity, and to include scarcity into the discussion resort must be made to economic theory.

IX

THE OIL WEAPON

"The scenario: an Arab embargo or supply cut, an atmosphere of crisis, probably in the aftermath of a short but bloody war. Then we go in."

MILES IGNOTUS, "SEIZING ARAB OIL," *HARPERS,* MARCH 1975.

We have seen that domestic energy problems originate in domestic political intervention into the free market process. Similarly, international energy problems originate, not surprisingly, from international political intervention. Certainly, if the governments of all countries stayed at home and looked after domestic business, we would have no conflicts and no disputes. However, an embargo or threat of an embargo on imported oil for a country heavily dependent on oil, such as the United States, is a potent weapon to be feared and without question can be used as an interventionist tactic likely to lead to greater conflict.

Yet our governing elite persists in manipulating world events into crisis. This manipulation has been well phrased by Colonel L. Fletcher Prouty, one time Director of Special Plans in the Office of the Joint Chiefs of Staff, who warned in his book *The Secret Team,* of "dangerous and illegal activities" carried on by the elite:

"The secret team responds to these emergencies. It responded to the petroleum shortage. Remember, the petroleum shortage supposedly came about when Saudi Arabia announced it would withhold petroleum from anyone who was friendly to the Israelis after the October 1973 war. All of a sudden our government decided we had a shortage of oil. But before that we didn't have a shortage of oil, and after a few months, when the price had been tripled, we didn't have a shortage of oil anymore. No new wells were dug. Nothing really happened except that was a 'response' to a 'crisis'...We had been getting less than 10 percent of our oil from Saudi Arabia. We increased imports from such places as Nigeria immediately and covered that 10 percent right away. Obviously, we had some concern over western Europe's demands, which is more dependent on Saudi Arabia. But the countries that supply the oil were under our control. The countries that are most under the control of the CIA are Saudi Arabia, Jordan, and Iran. Those are major oil supplier.(sic)"

Let us take a brief historical overview of these energy crises and petroleum shortages because of the October 1973 crisis cited by Colonel Prouty was not the first such crisis.

The first post World War II use of the potent oil embargo weapon was in the Suez crisis of 1956 which allied Britain, France and Israel against the Arab countries. In this crisis, Syria occupied Iraq Petroleum Company facilities, ARAMCO was forbidden to load tankers at Bahrein, the Iraq Petroleum Company pipeline to Haifa was blown up and no British or French tankers at all loaded Arab oil. Buried in news stories of the 1950s and 1960s we find a string of petroleum

TABLE 9-1: INTERRUPTIONS TO INTERNATIONAL OIL SUPPLIES FROM 1956 TO DATE

Arab countries	1956	Embargoed oil to France and Britain
Soviet Union	1956	Embargoed oil to Israel (equal to 40 percent of Israeli requirements)
Soviet Union	1958	Embargoed oil to Finland for political ends
Iraq	1966	Closed Iraq-Tripoli-Banias pipeline
Iraq	June 1967	Suspended pumping oil to Mediterranean terminals. Resumed operations for Syria, Lebanon, France and Turkey delivery on June 28 and for Spain on July 1. Embargo for U.K. and U.S. remained in force.
Kuwait	June 1967	Embargo on oil supplies for U.K. and U.S. (equal to 23 percent of U.K. requirements)
Libya	June 1967	Embargo on oil shipments. Exports for France, Turkey, Greece, Spain and Italy resumed July 3. Embargo retained for U.K. and U.S.
Bahrain, Qatar, Abu Dhabi and Saudi Arabia	June 1967	Embargoed oil for U.K. and U.S., ARAMCO resumed operations June 14 but for delivery to U.K. and U.S.
Syria	June 1967	Embargo on oil exports except for France and Turkey
Egypt	June 1967	Blocked Suez Canal with Egyptian ships
Algeria	1967	Nationalization of U.S., U.K. and Dutch oil companies, embargoed oil and natural gas to U.S. and U.K. (equal to 10 percent of U.K. natural gas requirements
Arab countries	1973	Embargo of oil to U.S. and Netherlands

embargoes since this 1956 incident. A summary of these supply interruptions can be found in Table 9-1.

The record shows that the Soviets used oil as a weapon against Israel long before the Arabs used the embargo against the U.S. On July 17, 1956 the U.S.S.R. signed a contract to supply the Israeli Fuel Corporation (Delek) with between $18 and $20 million of oil during a two year period — about 40 percent of Israel's oil requirements for 1957. Six months later, on February 7, 1957 the Soviets, abruptly and without warning, halted oil supplies to Israel just after deliveries had started. The Soviets then cancelled the two year supply contract citing Israel's "aggressive actions" against Egypt. The Israeli Foreign Ministry sent a note charging breach of contract and later Delek filed suit in Moscow for recovery of $2.3 million in damages. The oil contract covered a significant portion of Israel's requirements and was signed six months before supplies were breached. Without alternate supplies from the West, this single contract dependent on a hostile source could have spelled disaster to the Israelis.

THE 1973 EMBARGO

During the Six Day War of 1967, an Arab embargo was placed on oil to Great Britain, West Germany and the United States. But neither the 1957 nor the 1967 embargo was at all successful as a weapon. The Arabs continued their oil production and other producers around the world hastened to fill the supply gap; there was sufficient flexibility in the logistical system to dampen the effects of embargo.

The 1973 embargo reflected the earlier lessons and was probably somewhat more effective. Both direct and indirect shipments of petroleum were embargoed and Arab countries

TABLE 9-2: OIL PRICE COMPARISONS

	OLD PRICES (per barrel)	NEW PRICES (Dec. 1973 - per barrel)	
Iraq Iran Saudi Arabia Kuwait Abu Dhabi Qatar	$5.11	$11.65	effective January 1, 1974
Venezuela	$7.74	$14.08	Dec. 31, 1973
Indonesia	6.00	10.80	Dec. 31, 1973
Nigeria	8.31	14.60	Dec. 31, 1973
Libya	13.40	18.77	Dec. 31, 1973
Bolivia	7.44	16.00	Dec. 31, 1973

cut production. In mid-1973 Saudi Arabia, a United States client state, warned that continued oil supplies had become impossible because of the U.S. "complete support of Zionism against the Arabs." Following the outbreak of the fourth Arab-Israeli war, the Arabs used their oil weapon against supporters of Israel, not only embargoing to the U.S. and the Netherlands but simultaneously increasing prices. This had the effect of increasing Arab financial reserves and provided the economic means to maintain an embargo. On October 17, the Arabs increased oil prices by 17 percent, plus a 70 percent increase in taxes paid by the oil companies, and carried out the embargo threat. In October and November a series of production cuts were announced. The embargo was promptly seized upon by our elite, as Prouty indicates. Kissinger made a speech to the Pilgrim Society on December 12, 1973 to push for creation of an "energy action group of senior and prestigious individuals" to collaborate on a worldwide energy program. Thus, the oil crisis has been used from the start by the internationalists to foment a world order program rather than to solve energy problems. Effective January 1, 1974 the Arab countries doubled the price of oil and eased the embargo — but the damage had been done and the opportunity to create a crisis had been seized.

Naturally, the result of doubled prices was a dramatic rise in oil company profits. And the oil multi-nationals are dominated by those bankers and industrialists who make up the ruling elite, which Prouty accuses of fomenting the artificial crisis.

TABLE 9-3: NET THIRD QUARTER EARNINGS OF MAJOR U.S. OIL COMPANIES (July-Sept 1973)

MAJOR OIL COMPANY	EARNINGS 3rd QUARTER 1973 *	% INCREASE FROM SAME QUARTER 1972
EXXON	$638 million	+80 percent
TEXACO	307	48
MOBIL	231	64
GULF	210	91
STANDARD of INDIANA	147	37
SHELL	82	23
ATLANTIC RICHFIELD	60	16
PHILLIPS	54	43
CONTINENTAL	54	38
STANDARD of OHIO	18	14

SOURCE: *New York Times*, October 25, 1973.

Consequently, two significant observations emerge in the international oil play: political intervention by the United States in Mid East affairs leaves it exposed to Arab oil embargoes. At the same time major U.S. oil companies, highly influential in policy making through their elitist connections (the 'revolving door' and lobbyists) can reap handsome profits from oil embargoes.

It is of interest to note, in passing, the contributions made by these same multi-national oil companies to the Nixon re-election campaign just prior to the profitable 1973 embargo: Exxon contributed $442,000 (out of a profit of $1.6 billion in the first three-quarters of 1973) and Gulf Oil contributed $1,169,400 to the Nixon campaign (if we count Richard Scaife's $1 million). Ashland, Phillips, Occidental and others were also major campaign contributors.

THE OIL COST OF AN INTERVENTIONIST STANCE

Consequently, because the United States is an imperialist power with a seeming irresistable urge to create a New World Order, it must plan ahead for future embargoes and supply restrictions. A domestically-oriented stance would require only defense against external attack, an aggressive international stance requires more than coastal defense against economic counter moves.

The Hudson Institute has calculated what the United States must do in order to meet any future Arab oil restriction. Given a moderate planned build up for coal and oil shale facilities in the 1970s, a domestic demand of 1.5 million barrels of oil per day by 1980 and three million barrels per day by 1985, adequate energy defense would cost $20 billion for the decade 1975 to 1985 in addition to conservation measures.

The Carter Energy Plan tries to fulfill these requirements. Its tax clauses suggest that the American public is being called upon to finance a war plan under guise of a conservation measure.

The Hudson Institute also suggests that with adequate preparations, an Arab supply restriction would not be effective as a blackmail measure.[1] Planning for an Arab oil cutoff, as distinct from mere restriction, requires more heroic planning. The U.S. requires 18.3 million barrels a day but domestic production supplies only 9.8 million barrels. Imports from Canada and Latin America account for 2.3 million barrels and the Middle East supplies the balance of 6.2 million barrels. The current oil storage program is designed to give the United States time to set up rationing, increase domestic production and negotiate with other suppliers in the Soviet Union, Canada and Far East.

Even with high energy savings, such as conservation, reduction of dependence on Arab oil and oil storage, a considerable oil deficit will occur if Arab nations decided to cut off all oil shipments. The amount of this oil deficit has been calculated by the Hudson Institute under the two assumptions of moderate and high energy savings (Table 9-4).

OIL IMPORT QUOTAS

For several decades this security problem has been handled by an oil import quota system. This restricts foreign oil import and supposedly encourages domestic production. It has been widely asserted that the national security argument for oil import quotas is at least deficient and possibly non-existent.

(1) The Hudson Institute, Inc., Vol. I, *The Business Environment in 1975-1985*, (New York: 1974), pp. IV-22.

TABLE 9-4: OIL DEFICITS UNDER VARYING CONDITIONS

CONDITION	MILLION BBLS/DAY			
	1975	1980	1990	
A production restriction				
by Arab nations at 1974	.5	1.5	4.8	Moderate Energy Savings
supply levels	.2	.5	1.5	High Energy Savings
Arab nations cut off all	2.5	4.0	7.3	Moderate Energy Savings
oil	2.2	3.0	4.0	High Energy Savings

Source: The Hudson Institute, Inc., Vol. I, *The Business Environment in 1975-1985*, (New York: 1974) pp. IV-21.

Milton Friedman for example (*Newsweek*, June 26, 1957) argued "the political power of the oil industry, not national security, is the reason for the present subsidies to the industry. International disturbances simply offer a convenient excuse."

Senator Edward Kennedy, speaking from the standpoint of the high-cost Northeast made a parallel suggestion:

"Speaking as a New Englander, I have difficulty in contemplating the national security advantages under our present system in terms of delivery of oil to New England. I have difficulty in understanding whether it is going to make any difference at all, from the standpoint of national security, whether we import our oil from the Gulf ports and it comes by sea, or whether it comes from the Middle East or from Venezeula. If those ships are going to be torpedoed or sunk, it seems to me, our enemy is going to be able to sink a ship that is coming from the Gulf ports just as easily as it is from Venezuela. And I am just wondering on that point if my reasoning is legitimate or whether you find any real national security reasons for a continuation of the oil import program in terms of delivery of oil to the East Coast."[1]

(1) U.S. Senate, *Governmental Intervention in the Market Mechanism*, Part I, March 1969, pp. 23-4.

An earlier 1959 analysis, in the American Enterprise Institution series by W.H. Peterson,[1] also concluded that a "...study of the arguments used by the Government and the independent domestic oil and coal producers fails to find adequate cause to justify restrictions of oil imports," and then adds, "for these reasons, it is recommended that moves be instituted to dis-establish the entire quota program, whether voluntary or mandatory as expeditiously as possible." [2]

Given the Prouty argument we started with — that crises are used for the pecuniary advantage of an elite — then the oil import quota program criticized by Milton Friedman, Senator Kennedy and W.H. Peterson, may well be a method of shunting profits to more politically activist multi-nationals.

Leaving this intriguing possibility to one side, the empirical evidence does suggest that the Soviet Union and Middle East states are well aware of, have used, and would presumably again use, oil as a weapon to achieve national objectives. Examples cited above include Soviet interruption of crude oil for Israel (amounting to 40 percent of Israel's requirements); Kuwait interruption of crude oil for the United Kingdom (amounting to 23 percent of U.K. imports and Algerian interruption of natural gas for the U.K. (amounting to ten percent of U.K. natural gas imports). These are not insignificant inconveniences. They are serious and deliberate interruptions in the supply of a vital commodity to achieve a national objective. We also find that the U.S. is directly and indirectly dependent on petroleum sources subject to such an interruption of supply. Further, European allies of the U.S. are critically dependent on crude oil and natural gas from sources

(1) William H. Peterson, *The Quest of Governmental Oil Import Restrictions*, (American Enterprise Institute, 1959).

(2) *Ibid.*, p. 67.

subject to interruption of supply, even more so than the U.S. The burden of solving the problem of any interruption would fall heavily upon the United States. It is therefore clear that the Free World is in a potentially precarious position, if the U.S. adopts an interventionist foreign policy against oil producers.

Government restrictions on the other hand, have limited the amount of domestic crude produced. Both drilling and finding rates have declined substantially in the 1970s, and to solve this national security problem, Project Independence was established.

PROJECT INDEPENDENCE OR WAR ALERT?

Oil and gas independence for the U.S. is feasible because about one-half of original recoverable resources still remain in the ground, and even larger areas remain to be explored and developed.

Certainly world oil and gas is not nearing depletion, and this suggests competition for Arab producers sometime in the future. In early 1978, Venezuela announced that its under-developed Orinoco Basin reserves contain not less than 700 billion barrels — or about twice the known reserves of Saudi Arabia. Mexico has equally large reserves.

Project Independence ignores these close-to-home sources and is the focus of a plan with the declared purpose "to satisfy energy demand at minimal cost without dependence on insecure foreign sources." Notably, Project Independence emphasizes government planning and reduction of demand through conservation. The reports are based on an input-output computerized model converted into a so-called national energy plan. Critics who pointed out there is no energy shortage and

TABLE 9-5: WORLD & U.S. OIL AND GAS PRODUCTION, RESERVES AND RESOURCES (as a percentage of original recoverable resources)

	REMAINING RECOVERABLE RESOURCE (12/31/75)		PROVED RESERVES (12/31/75)		PRODUCTION 1976	
	U.S.	World	U.S.	World	U.S.	World
Oil	56.4	83	13.1	29.3	1.18	1.02
Natural Gas	64.4	92	16.3	21.8	1.41	0.46

therefore no need for a plan are ignored. Those who point to the benefits of such a plan accruing to an elitist group are silenced, although Nelson Rockefeller's $100 billion energy boondoggle has been squashed. The thrust of national planning has not been halted, but is buried within Carter's energy plan.

X

THE CARTER ENERGY-TAX PLAN

"Ninety-six per cent of Americans will see their taxes go down...."

PRESIDENT JAMES EARL CARTER, STATE OF THE UNION MESSAGE, 1978.

President Carter has declared the "moral equivalent of war" upon our energy crisis, a declaration that reminds one of Friedrich Hayek's pertinent observation on collectivism and morality: collectivist programs which spring from supposed high moral motives in practice lead to a system where the "worst get on top."[1]

The Carter imperialist and some of the oil corporate collectivists (the *Wall Street Journal* cites Atlantic Richfield as

(1) Friedrich A. Hayek, *The Road to Serfdom*, (Chicago: University of Chicago Press, 1944), p. 136.

one spokesman for this group),[1] are in partnership to create an artificial belief structure in the American people as far as an energy crisis is concerned. The main elements of this belief structure appear to be:

(a) that we have an absolute shortage of energy resources, and

(b) that the only solution to this crisis is to reduce our consumption of energy by government control.

Once again Hayek, in *The Road to Serfdom*, reminds us that a collectivist community takes great interest in the consumption habits of its individual members, encouraging some habits, while discouraging others, to bring about the desired new order.[2] Of course, no individual can be allowed to retain objectives which stand in the way of elitist goals. When we look at the Carter Energy Plan in terms of a collectivist framework and a synthetic belief structure, we can explain its uselessness as an energy plan and also note its potent ability to order behavioral changes in the American community by taking from some and giving to others.

THE CARTER PLAN AS SOCIAL ENGINEERING

If the Carter Energy Plan can be foisted on Congress (and in great part it probably will be), it proposes to do the following:

(a) raise the cost of energy to the consumer by taxation, that is, it's a tax program rather than an energy supply program.

(b) rebate part of the tax to favored recipients, selected on

(1) *Wall Street Journal*, August 11, 1977.

(2) *Op. cit.*, p. 147.

political grounds,

(c) grant some commercial consumers a selective program of tax credits,

(d) use this program of taxation to restrict consumption and ignore production,

(e) hand out Federal grants as incentives for such restriction of end uses.

The Carter energy program uses a synthetic energy crisis as a vehicle to bring about structural changes in American society. The emphasis is not upon production to solve a shortage. For instance, nuclear energy, the cheapest and most viable long-run energy, is dreaded by the Carter Administration. Moreover, the emphasis is to cut energy use, and thus force down the American standard of living and redistribute wealth by selectively taking from some and giving to others. The Carter Energy Plan is a people control device, not an energy supply device.

To do this, the Carter energy-tax scheme proposes the following specifics:

(1) impose a graduated excise tax on new automobiles and light trucks unable to meet economy fuel standards, with graduated rebates for vehicles with better than standard fuel economy. These taxes would be gradually increased each year through 1985. These fuel economy standards are meaningless: miles per gallon figures issued by the Federal government are hopelessly optimistic and arbitrary. Thus, it will be impossible for industry to meet knowable objective standards. At any time on a whim, simply by calling for accurate standards, the Federal bureaucrats can immobilize the automobile industry.

(2) A standby gasoline tax is to be imposed if national gasoline consumption exceeds an unstated, yet to be determined, target. This revenue is proposed to be rebated through direct payments to those people who do not pay the tax. Again, this is a political device to take from some to give to others and build political capital in the process.

(3) Removal of the ten percent excise tax on inter-city buses. Why? To get people out of cars, out of airplanes and into buses.
(4) Elimination of Federal excise tax preferences for general aviation and motorboat fuel. Why? To get people out of small airplanes and motorboats.
(5) An oil equalization tax equal to the difference between the government controlled price of domestic oil and the world price of oil, supposedly to be rebated to consumers.
(6) Every user of natural gas (except manufacturers of fertilizers) is to be taxed an amount equivalent to the difference between the average cost of natural gas and a target price determined by the cost of oil. This is a simple tax device and has nothing to do with new energy supply.
(7) A tax on individual consumers of petroleum products equal to 90 cents (rising to $3.00) a barrel.

The Carter social engineers propose to take this flow of energy consumer funds and channel it back to other consumers but in different amounts and directions to achieve social objectives:

(1) Rebates for motor vehicles with better than standard fuel economy, i.e., to get people into small automobiles,
(2) Rebates from the gasoline tax through the general tax system,
(3) A tax credit for home owners equal to a percentage spent on weatherization and installation of solar equipment,
(4) Tax credit for business for investment in approved conservation measures, and a credit for purchase of "cogeneration equipment,"
(5) Tax benefits for industrial users for conversion from oil to coal and natural gas. Presumably this does not exclude industrial users who have already received credits for switching the other way — from coal to oil under yet another government program,
(6) Production incentives for independent gas and oil drillers and geothermal well drillers; presumably to allow them to carry the burden of below-market clearing prices for oil and gas,

The energy plan is an incredibly complex piece of tax legislation designed to raise taxes, reduce overall energy consumption, and divert industrial energy use from natural gas and oil to coal. As a design to combat a mythical day in 1990 or 2000 when the world abruptly runs out of gas and oil, it is difficult to determine the relationship between the plan and any energy shortage. Even the Comptroller General's Office, which has argued "on the basis of our prior work," that there is a "serious energy problem," comments that the Carter plan cannot be achieved without either unspecified voluntary efforts, or unspecified mandatory controls. Even if the plan is implemented, and Congress passes it, Administration goals cannot be achieved:

CARTER STATED ENERGY GOALS FOR 1985	ADMINISTRATION'S ESTIMATE OF WHAT THE CARTER PLAN WILL DO BY 1985
1. Reduce total energy growth to below two percent per year	Reduction to 2.2 percent
2. Reduce oil imports below six million barrels per day	Reduction to seven million barrels per day
3. Insulate 90 percent of all buildings	Insulate approximately 60 percent of all buildings
4. Use solar energy in 2.5 million homes	Use solar energy in 1.3 million homes

If the Carter Plan won't achieve the Administration energy goals it certainly will redistribute income, at first taking from some to give to others; then, sometime before 1985, taking from all for more bureaucratic waste.

DISCRIMINATORY EFFECTS OF CARTER'S ENERGY TAX PROGRAM

Fundamentally, the Carter energy program has little to do with using abundant American energy reserves, but has a great deal to do with re-allocating income and generally depressing the American standard of living for no publically ascertainable purpose. As an energy program, it is deception. But how much so? The effects of the Carter energy tax and credit proposals in 1980 are as follows:

TABLE 10-1: TAXES AND CREDITS IN THE CARTER ENERGY PLAN

INDIVIDUAL INCOME	ENERGY TAXES ON INDIVIDUALS		ENERGY CREDITS FOR INDIVIDUALS		NET CHANGE IN REVENUE	
	$ millions	percent distri-bution	$ millions	percent distri-bution	$ millions	percent change
Less than $15,000	+5,341	48.9	−6,923	70.9	−1,582	−3.8
$15,000 to $30,000	+4,048	37.1	−2,319	23.8	+1.729	+1.9
Over $30,000	+1,522	13.9	−520	5.3	+1,002	+1.2
TOTAL	10, 913	99.9	−9,766	100.0	1,147	0.5

Source: U.S. House of Representatives, Hearings before the Committee on Ways and Means, *Tax Aspects of President Carter's Energy Program*, 95th Congress, 1st Session, May 16-19, (Washington, D.C.: 1977) pp. 172-73.

The total revenue effect of the Carter energy program is to raise $10.9 billion of new taxes by 1980 and rebate just under

$9.7 billion to selected energy users, with a net gain for the Federal government of $1.1 billion. The most obvious discriminatory effect is to provide Federal handouts to those earning less than $15,000 a year: this group gains 3.8 percent in net revenue terms. The plan also takes from those earning over $15,000 a year: the net revenue effect in these groups 1.9 percent ($15,000-$30,000 bracket per year) and 1.2 percent (over $30,000). The long suffering middle class will pay $5.5 billion in new taxes and only receive rebates of $2.8 billion. The energy tax plan is another device to eliminate the middle class in America.

TABLE 10-2: THE IMPACT OF THE CARTER ENERGY TAX PLAN IN 1985

	ENERGY TAXES ON INDIVIDUALS		ENERGY CREDITS FOR INDIVIDUALS		NET CHANGE	
Income	$ million	percent	$ million	percent	$ million	percent
Less than $15,000	8,960	47.4	−7.126	72.0	+1,834	2.5
$15,000 to $30,000	7,000	37.1	−2,259	22.8	+4,741	3.0
Over $30,000	2,934	15.5	−510	5.2	+2.424	1.7
TOTAL	18,893	100.0	−9,894	100.0	+8,999	

Source: *Ibid.*

Moreover, the rebate procedure is illusory even for the favored group earning less than $15,000 a year. Persistent ad unsolvable inflation drives workers into brackets above $15,000 a year — the poverty level threshold of the 1980s. So all income groups will find, by 1980, that the so-called energy program is a tax program and the rebates an illusory carrot eaten up by price inflation.

And don't forget the key point — this energy tax program does precisely nothing to increase production of energy. The plan attempts only to restrict consumption and does this in a deceitful and illusory manner.

Opposition to the Carter plan has ranged the political spectrum. The NAACP, representing low income black workers, dislikes the concept of using energy to raise taxes, and recognizes this will have no effect in increasing the supply of energy. The NAACP did not, however, focus on the illusory nature of the so-called benefits intended to appeal to low income workers to get the program through Congress. These benefits will become net liabilities by 1985.

TABLE 10-3: THE DISAPPEARING CARTER CARROT: NET CHANGE IN ENERGY TAX BY 1985

1973	−$ 801,000,000 credit to low income workers
1977	+$ 298,000,000 extra tax
1980	+$1,147,000,000 extra tax
1985	+$8,999,000,000 extra tax will include low income workers

Source: As Table 10-1.

The NAACP made a strongly worded critique including the following comments, in line with the fundamental argument of this book:

"We have examined the Administration's National Energy Plan in the light of the agenda for economic growth and development for America's Black people. What we see in the plan is an emphasis on conservation, and a reduction in the growth of total energy demand and consumption.... This emphasis cannot satisfy the fundamental

requirements of a society of expanding economic opportunities.
"We think there must be a more vigorous approach to supply expansion and to the development of new supply technologies so that energy...can continue to expedite economic growth and development in the future. All alternative energy sources should be developed and utilized. Nuclear power, including the breeder, must be vigorously pursued because it will be an essential part of the total fuel mix necessary to sustain an expanding economy. Other alternative sources...must also be developed and made commercially available at the earliest possible time."

and,

"...We are fearful that an energy policy with an overriding concern for protection of the environment may cause governmental policy-makers in this area to lose sight of the other more compelling economic and social objectives....
"We recognize that nuclear power does present certain problems. But we think these problems can be solved through dedicated efforts by government, the scientific community, and industry working cooperatively together. Notwithstanding the claims of opponents of this source of energy, the fact is that nuclear power will be required to meet our future needs for electricity..."

The *Wall Street Journal* has been equally critical of Carter's tax plan:

"As the Senate returns from its recess, it finds on its desk the largest peacetime tax increase in the nation's history. Mr. Carter calls his tax boost an 'energy program,' but in fact it is a cleverly disguised grab for the nation's paychecks."[1]

The *Journal* calculation is that,

"The Carter program would increase taxes by well over $20 billion,

(1) *Wall Street Journal,* September 7, 1977.

and perhaps more than $100 billion if the administration suceeds in its attempts to revive the gasoline tax."

In a series of related articles, the *Wall Street Journal* attacked Carter in editorials with such titles as:

"1,001 YEARS OF NATURAL GAS" (Apr. 27, 1977)
"ERDAgate" (May 20, 1977) and
"The Energy Crisis Explained" (May 27, 1977.

The last editorial began with these words:

"The energy crisis is a snare and a delusion. Worse, it's a hustle."

The *WSJ* editorials state that we are nowhere near consuming all the planet's oil and natural gas. All we have to do is get the politicians out of the picture and let supply and demand work it out in the marketplace.

A BUREAUCRACY FOR CREATED CRISIS

The Carter created crisis is more than just a tax plan to re-distribute income. It requires a new Cabinet-level department — the Department of Energy (combining the former Federal Energy Administration, the Energy Research and Development Administration and the Federal Power Commission under one bureaucratic roof) to maintain the crisis and centralize control.

This department came into operation on August 4, 1976 with James Schlesinger as Secretary, a $20 billion budget and 20,000 bureaucrats. However, proposals for the Department of Energy were promoted long before this. President Richard Nixon in 1973 proposed legislation to establish a Department of Energy and Natural Resources with functions transferred from the Department of the Interior and several other agencies, to appoint an Assistant to the President for Energy to head the Energy Policy Office; establish a new independent agency, the Energy Research and Development Administration (ERDA), to focus all Federal energy research and development, and retain the five-member organization of the Atomic Energy Commission for licensing, regulatory and related environmental and safety functions but under a new name, the Nuclear Energy Commission.

The Republican Nixon Energy Department was similar to Carter's in that it was:

"responsible for the balanced utilization and conservation of our Nation's energy and natural resources. The Department would bring together and realign many related Federal programs which are now scattered among several departments and agencies. DENR would have the responsibility for assuring that future demands for water, timber, minerals and energy resources are met without sacrificing our forests, lakes, wilderness, beaches and the general environment. It would foster a better understanding of the total environment — the oceans, the atmosphere, the lands and their interaction."

To do this, Nixon's concept was as socialist as Carter's DOE.

"would have an organization and managerial capability which could most effectively and vigorously develop and implement comprehensive natural resources policies and programs."

The Schlesinger personality is an interesting indicator of the direction of the Carter version of the Nixon energy plan. The *Wall Street Journal* (July 8, 1977) ran an excellent profile describing Schlesinger as "a deep thinker" who "lacks something as a salesman or a legislative technician." Formerly with

RAND Corporation in Santa Monica, Schlesinger has also been head of the Atomic Energy Commission, the CIA, and was Defense Secretary until fired by President Ford. According to his wife (reports the *Journal*), "he has few outside interests except for bird watching (he says he has identified more than 600 species)," and because of his gruff and arrogant manner "he doesn't have any friends, only followers." More disturbing, in light of our thesis of a trend towards the American dictatorship, in this self-portrait reported by wife Rachel, "They're saying I'm arrogant. They're right. Goddamn it, I am arrogant, I can't stand stupid questions."

Schlesinger's autocratic nature is well known to President Carter and is reflected in the official press release on Schlesinger's appointment. In this, the President remarked as follows:

"Of course, the next problem that we face is the selection of a person to head up the new Department of Energy. This has been a matter that has been of great concern to me for the last few months. I have decided to establish a search committee — (Laughter) — to choose a Secretary, and I have asked Dr. James Schlesinger to head up the search committee. And at his request, the membership of the committee will be limited to one person. (Laughter)."[1]

This arrogance explains why, within a month of taking office, the Carter energy team floated the idea of a Youth Energy Program. Under this program the state enrolls youths and arms them with a questionnaire and rule book, to knock on doors. The young inspectors, presumably upon invitation, then comb houses for energy problems, everything from checking the air in the car's tires to inspecting the level of water in the toilet tanks.

(1) Office of the White House Press Secretary, August 4, 1977.

HEIL!
WE ARE FROM THE CARTER ADMINISTRATION NEIGHBORHOOD YOUTH ENERGY CONSERVATION INSPECTION BROWN SHIRTS!
WE HAVE COME TO INSPECT YOUR INSULATION!
YOU PROBABLY THINK YOU'RE DREAMING
EVERYTHING WILL BE OK.. IF YOU OBEY ORDERS'

The Nazi state mentality behind this proposal quickly penetrated the media, and generated a classic cartoon of the Carter Energy Youth Corps replete with swastika arm bands. Schlesinger's Big Brother mentality and the philosophy behind his initial proposal should not be ignored, it is a surfacing of the totalitarian philosophy.

The Youth Corps trial balloon was followed in December 1977 by reports of suppression of already published official information that disproved the Carter thesis that an energy crisis was imminent. With the approval of Secretary Schlesinger, the Office of Energy Information and Analysis manipulated statistics in an official report so as to increase estimated energy demand by the equivalent of 2.52 million barrels of oil a day by 1985 and decrease anticipated supply by 1.44 million barrels of oil a day by the same date. This artificially created a deficit of 3.96 million barrels of oil per day.[1]

In spring 1978, Big Brother proclivities of Schlesinger's DOE surfaced once again. A DOE report, which accurately reported ample natural gas resources in the United States was purged and recalled as containing "erroneous information." Indeed the report was erroneous in the sense that our ample methane resources fail to support a created crisis scenario. The DOE required the U.S. Government Printing Office (GPO) to destroy all the offending reports. The GPO did so in the following letter to its depository librarians:

"ATTENTION
DEPOSITORY LIBRARIANS:

The Department of Energy has advised this office that the publication *Market Oriented Programs Planning Study* (MOPPS), *Integrated Summary Vol 1, Final Report, December 1977*, should be removed from your shelves and destroyed. The publication was shipped on S/L

(1) *San Jose Mercury*, December 21, 1977

10,558 (2nd shipment of February 7, 1978), under Item Number 429-P (El.18.0011/1 (D). We are advised the document contains erroneous information and is being revised. Your assistance is appreciated.

J.D. LIVSEY,
Director, Library and Statutory
Distribution Service (SL)

U.S. Government Printing Office
Washington, D.C. 20401."

How did a single copy of the offending report escape the Big Brother tentacles of the DOE watchdogs? Apparently Paul Schaffer, editor of *Energy User News* spotted an advance notice of the MOPPS study and spent a Saturday "in the Newark Public Library feeding 376 dimes into the photostat machine."

The pity of it all is that howls of indignation have so far been too few. Apparently, Congress does not see, or refuses to see, the 1984 monster it has created in the name of an "energy crisis." But the *Wall Street Journal* summed up Carter's policy succinctly:

"Intellectually and morally, the Carter energy program is bankrupt, and is increasingly recognized as such in those quarters of opinion that have a conception of the common good. Politically it survives and perhaps prospers. It has passed the House, and large parts of it may become law, but the only remaining force behind it is the crassest sort of political muscle."[2]

(1) *Wall Street Journal,* April 7, 1978. Also see April 4, 1978 editorial, "The Memory Hole."

(2) *Wall Street Journal,* August 1, 1977.

Chapter XI

XI

THE WASHINGTON ENERGY HUSTLERS

"The energy crisis is a snare and a delusion. Worse, it's a hustle.[1] We're now prepared to explain it for once and for all."

EDITORIAL, *WALL STREET JOURNAL*, MAY 27, 1977

From one end of the political spectrum to the other, among those without energy-related special interests, there is a general belief that the energy crisis does not exist. It's an artificial belief structure, a con game, or as the august *Wall Street Journal* phrases it, "a hustle." The energy crisis is indeed a created crisis.

(1) "HUSTLE," v. to make money in any illegal enterprise; to seek customers for, or as, a prostitue; to seek customers for a rigged game of chance or other confidence game — hustler, n." *A Dictionary of Contemporary and Colloquial Usage*, p. 33.

It won't be news to many that hustlers occupy the luxuriously carpeted suites along Constitution Avenue, many taxpayers have long believed such notions. What is news today is the wide range of people believing in this view. When the U.S. Labor Party's *New Solidarity* makes the same critical points as the *Wall Street Journal,* and the National Association for the Advancement of Colored People (NAACP) makes the same arguments as the white supremist *Spotlight,* then something strange is happening indeed.

A general overview of the nature of the created crisis can be gleaned from Part One of this book where we described the domestic energy sources of the United States. There is no absolute shortage. We have abailable now at least 2,000 years of energy resources. True, we will have to develop new energy technologies, and this development is well within the scope of private initiative if we allow the inventors, the innovators and the private business mechanism to get on with the job.

Then in Part Two we outlined reasons why the shortages are created, and they ultimately can be reduced to a single word — politics. At first, political decision making replaced economic decision making in, for example, the Natural Gas Act of 1938 and the Supreme Court Phillips Petroleum decision of 1954. This produced an artificial shortage of natural gas. A political price is always different from a market price; condsequently political prices always bring about either surpluses (hospital beds) or shortages (natural gas). Political attempts to correct the problem suceed only in compounding the disaster.

Today political intervention into the market process is supplemented by political intervention into the technological decision making area. Politicians have become instant engineers as well as instant economists. Governor Brown of California has decided that solar energy is the cure. President Carter is flat out against breeder reactors. The CIA makes a study to show the Soviets are short of oil, so the study is released. Energy Secretary Schlesinger doesn't like geological calculations so the geologists are censored. Technical choice must be a function of engineering data and economic calculation in the same way that price is determined by the market. In no way can a politician do more than take a wild guess at efficient technical choices. So we now have a society whose energy supply is being determined by wild guesses dependent on ideological, sociological or personal aesthetic considerations and energy prices by arbitrary social engineering standards.

The crises are indeed the effects of politics of "planned chaos," as Ludwig von Mises has phrased it, is verified by two articles in the *Monthly Review* of the Federal Reserve Bank of St. Louis, cited not because this is a better or worse source than a hundred others, but because the FRB is a quasi-official body. In the November 1975 St. Louis *FRB Monthly Review,* an article by Hans H. Helbling and James E. Turley entitled "Oil Price Controls: A Counterproductive Effort" refutes the price controls imposed in August 1971. What was the result of the controls? It was the opposite of that promoted by the politicians. Controls discouraged production of oil, and were incentives to import oil and ultimately assisted rising oil prices. All these are, presumably, the exact opposite to the political intention. And remember that big business collectivists such as Thornton F. Bradshaw, President of Atlantic Richfield Oil Company want government controlled oil prices and have publically stated this objective. Their enthusiasm is understandable.

Another article in the same publication (May 1977) by Robert H. Rasche and John A. Tatom was entitled "The effects of the new energy regime on economic capacity, production and prices." The quadrupling of oil prices in 1973, stemming from U.S. intervention (or an oil company conspiracy if you wish) also had the effect of permanently reducing U.S. economic capacity by 4 to 5 percent, reducing the productivity of labor and any attempt to close the gap (estimated at $100 billion) would stimulate price inflation.

So in effect, the political tail wags the technio-economic dog. The only way we can stop the planned chaos — because planned it obviously is — is to remove energy decision making power from the manipulators in Washington.[1] and the big business social engineers who prefer statist planning to a free market economy.

But this political surgery is beyond human powers because the grass roots activist proponents of planned chaos, almost without exception, display a lack of even elementary understanding of the adverse consequences of their planning. They are not open to conversion by logic and reason. Not one proponent of planned energy has given equal attention to the cost of planning. The advantages of a particular planning mode are always presented — but never the costs. Or, if they do understand the consequences, they are promoting a terrible theft from the American consumer.

Possibly the only way the substitution of special-interest ideology for economic and engineering rationality can be reversed is by emphasizing the dollar cost of ideology to the general public. Look, for instance, at the following figures on the savings from nuclear power generation.

Far from nuclear power being a monster to be avoided, the cost savings of nuclear power are such that it should be welcomed with open arms. After all, a $20 to $50 per year reduction in a residential electricity bill is more beneficial to all income groups than the energy credit tax bill proposed by the Administration. We have already disposed of nuclear scare tactics; coal mine and pipeline accidents are many times more

(1) An interesting article along these lines is *The Case for Political Action* by Howard S. Katz, *Reason*, (April 1975).

TABLE 11-1: DOLLAR SAVINGS FROM NUCLEAR GENERATION OF ELECTRICITY AT GENERATING STATIONS USING NUCLEAR AND COAL/OIL FUELS

UTILITY	NUCLEAR PART OF GENERATING CAPACITY	DOLLAR SAVINGS	
		Total to customers (Millions $)	Typical residential customer
Baltimore G&E	39.8%	$26.9	$38.28
Commonwealth Edison	40.0%	184.1	$20.49
Duke Power	25.0%	61.0	$54.87
Philadelphia Elect.	25.8%	66.0	$15.12
Southern California Edison	4.1%	38.6	$13.72

Source: Atomic Industrial Forum.

numerous than nuclear accidents. Indeed there is no real comparison because there have been no nuclear power plant fatal accidents.

Yet because the energy industry has long used the political process to enrich itself, the industry is no longer credible to the public. Congress opposes the oil multinationals, threatening divestiture and nationalization. The investigation of Gulf and the international uranium cartel was a field day for publicity minded Congressmen:

"Says McAfee, Chairman of Gulf Oil, sadly: Frankly, I would characterize it as a kangaroo court. It was very revealing in recognizing what we are up against."[1]

(1) *Forbes*, October 15, 1977, p. 95.

The industry will find little public support in any attack from Congressional quarters because historically petroleum lobbyists have lingered too long at the public trough.

Without question, the most powerful opponent of free markets in energy (and only free markets will provide the solution) are to be found within big business — those neanderthal executives who talk free enterprise in public and promote socialism in private. A prime example we have cited is Thornton F. Bradshaw, president of Atlantic Richfield — whose pro-Big Brother government views probably stem from his tenure as a Harvard School of Business professor. Thornton loves big government. He considers Arab oil prices too moderate. He wants, "the permanent management of crude oil prices by the U.S. government." Fixed oil price proponents like Thornton lend some credence to the theory that Arab quadrupling of oil prices in 1974 was rigged by multi-nationals and their New York bank stockholders. After all, the New York banks gained most from oil price increases: they act as a funnel for the new flow of funds, they float loans for LDCs unable to pay increased oil bills while other U.S. companies under their control gain from the massive Arab orders for American technology and goods. Of the $194 billion surplus accumulated by OPEC between 1974 and 1978, just under $50 billion has been invested in the United States, in property, in U.S. government securities and in the stock market. The fourfold oil price increase was a gigantic income transfer from the pockets of oil consumers to the pockets of those banks and multi-nationals (and of course their workers) catering to Arab needs.

Certainly if Congress wants to investigate an energy problem it could well start with the role of the New York banks in the 1973 OPEC embargo. Moreover, when one finds that funding for noisy anti-nuclear forces comes directly or indirectly from foundations associated with these multi-nationals we have reason indeed to wonder...and perhaps be more than a little suspicious.

When we examine the views of Establishment think tanks and publications — those that give the clue to the policy road ahead — our conclusions are darkened even more. In the October 1976 issue of *Foreign Affairs,* that quarterly Establishment oracle which reflects elitist thinking and intended action, we find an article by Amory B. Lovins. Lovins is not a household word and has hardly made a scratch on the world of knowledge and ideas. However, merely because *Foreign Affairs* is his forum, Lovins' ideas became acceptable, are widely reported and taken seriously by high officials. Lovins is an instant engineer. His aim is to substitute "soft" technology for hard technology, that is, windmills and solar panels for electric power stations and synthetic gas plants, all without a hint of economic or technical analysis. In fact, Lovins writes as if analysis does not exist. One commentator summed it up very well:

"Much of Mr. Lovins' theory is reminiscent of certain ideas utilized by the People's Republic of China during the years of the Great Leap Forward."[1]

A brief excerpt from Lovins will surface the approach:

"If you ask me, it'd be little short of disastrous for us to discover a source of clean, cheap, abundant energy because of what we would do with it..."

A concluding note to the saga of an artificial crisis created by ideologies comes from a recent (February 1978) editorial in *Canadian Petroleum*, after an editorial tour of Canadian energy producing areas. The editor outlines the current boom in Canadian drilling operations, the search for oil

(1) Daniel W. Kane, Council on Energy Independence, Chicago.

and gas, the labor shortages and equipment shortages and the manner these are solved on a day to day basis by private enterprise without resort to government planning and crash solutions. Then the *Canadian Petroleum* editorial spots a cloud in the sky:

"That's the good news. The only dark cloud that everyone I talked with saw on the horizon, was that some idiot politician might put his foot in his mouth...it's happened so many times in the past that people are still a little gun shy."

AFTERWORD

The manuscript of *Energy: The Created Crisis* was written in early 1978. In October 1978, the Carter Administration finally wrested its long awaited energy bill from a squabbling Congress. The legislation emphatically confirms the argument of this book: the energy crisis is a phony, a "created crisis" to be "managed" by a self-appointed elite.

Furthermore, in the opinion of many independent observers outside Washington, the Carter legislation will do little to save either oil, gas or dollars; quite the reverse in fact. The new energy program is so complex and unwieldy that bureaucratic bumbling will make equitable or efficient administration almost impossible. Rather than deregulate — as the administration has done so successfully with the airline industry — the legislation adds a thicket of new regulations based on geological and geographical subtleties.

Take natural gas, for example; Price limits on most newly discovered gas are to be removed, but only in 1985 and replaced long before 1985 by an extraordinarily complex regulatory system that is virtually guaranteed to restrict gas development. Whereas regulated gas prices today are based only on the date of start-up of the well, prices under the new Carter system will be based on a maze of variable geological characteristics and the history of each individual well. The system of natural gas pricing has not been simplified. On the contrary, it has been made far more archaic and inefficient. The result will be less gas — a "created crisis" ready to be "managed" by our elitist professional crisis managers.

Former Federal Power Commissioner Connole has commented that the new legislation is "...an almost bottomless pit of opportunity for argument and contention" and "they'll be ten years of litigation on it...."

Adds Mr. Flug, Director of the consumer group Energy Action, "the new gas classification system is an invitation to fraud and abuse."

For what purpose? Industry contends there will be an increase in energy production. For example, Alan Cope of Continental Oil says:

"The gas bill is so complex and so awful in its administrative aspects that it's really questionable that it will produce any more gas at all than the present regulatory system."

Which brings us back to the basic argument of *Energy: The Created Crisis*. The energy "crisis" is a phony, a rip-off, a political con game designed to perpetuate a "crisis" that can be "managed" for political power purposes.

A.C.S.

GLOSSARY

AEC	Atomic Energy Commission
AIF	Atomic Industrial Forum
Chain Reaction	A self-sustaining sequence of nuclear fissions taking place in a reactor core.
Combustion	Burning of fossil fuels to release chemical energy.
Containment	The structure, usually of reinforced concrete, designed to isolate fission products from the environment in the event of a major nuclear accident.
Control Rod	A neutron absorber, used for control of the chain reaction by insertion or withdrawal in the core.
Core (reactor)	The central portion of a nuclear reactor containing nuclear fuel, moderator and control mechanisms.
CRBR	Clinch River Breeder Reactor. A demonstration power-breeder, near Oak Ridge, Tennessee..
Decay	Process of radioactive disintegration.
Deuterium	A heavier, naturally occurring isotope of hydrogen. In nature, it forms 0.016 percent of hydrogen and is a potential fusion fuel.
DOE	Department of Energy.
ECCS	Emergency Core Cooling System. A reactor safeguard to return coolant to the reactor core in the event of a loss of coolant.
Enrichment	A process, either gaseous diffusion or centrifugation, whereby the U_{235} content of uranium is increased.

Environmental Impact Statement (EIS)	A statement setting forth the probable environmental consequences of building and operating facilities
EPA	Environmental Protection Agency.
EPRI	Electric Power Research Institute.
ERDA	Energy Research and Development Administration.
ERDA-48	A publication of the Energy Research and Development Administration, titled "A National Plan for Energy Research, Development and Demonstration." June 1975.
Fission	Nuclear process in which a heavy atom is split into fragments.
Fission Products	Atomic fragments created by nuclear fission: includes more than 30 elements and several hundred nuclear isotopes.
Fuel Reprocessing	A chemical operation in which nuclear fuel is recovered from spent fuel.
Fuel Rods	Tubes of zirconium alloy containing uranium oxide fuel pellets.
Fusion	A nuclear process in which two light atoms fuse to form helium.
GEC	Gross Energy Consumption
Geopower	Also geophysical power. Power derived from geophysical sources such as wind, tides, sun and ocean.
Heavy Hydrogen	Hydrogen found in nature having twice the weight of ordinary hydrogen or protium.
Hydrocarbon	A chemical compound or molecule composed of carbon and hydrogen atoms.
Hydropower	Also hydroelectric power. Power generated from falling water.

JCAE	Joint Committee on Atomic Energy. A special committee of Congress for atomic energy matters.
Kilowatt	kw, 1000 watts of power.
Laser	High intensity coherent radiation. (Light Amplification by Stimulated Emission of Radiation.
LMFBR	Liquid Metal Fast Breeder Reactor.
LNG	Liquified natural gas. At minus 160 degrees C, natural gas becomes a liquid.
LWR	Light Water Reactor.
Mcf	One thousand cubic feet, a unit for natural gas volume.
Megawatt	Mw. One million watts; 1,000 kilowatts. Mwe refers to a megawatt of electrical power.
Methane Methanol	Wood alcohol. (CH_3OH)
Metric ton	2205 pounds.
Mev	One million electron volts.
MHD	Magnetohydrodynamic. A process in which energy is extracted from high temperature ionized gas as it passes through a magnetic field.
Mwe	Megawatt electrical. 1,000 kilowatts.
NASA	National Aeronautics and Space Administration.
Natural gas	Methane (CH_4) and hydrocarbons found in gas wells or associated with oil.
Neutron	A fundamental particle having no electrical charge.
NRC	Nuclear Regulatory Commission. A federal agency created in 1974 to regulate the nuclear power industry.

NSF	National Science Foundation
Nuclear power	The controlled release of fission power in a nuclear reactor.
Pellet	Uranium oxide pellet stacked inside a zircaloy sheath to form the nuclear fuel rod.
Petroleum	Oil liquid hydrocarbons.
Photoenergy	Energy from conversion of radiant (solar) energy.
Plutonium	Element 94, chemical symbol Pu. Plutonium-239 is formed by neutron absorption in uranium-238.
Pressure vessel	A thick-walled steel vessel to hold the reactor core.
Price-Anderson	An Act of Congress limiting liability in the event of a nuclear accident with release of radioactivity.
PWR	Pressurized Water Reactor. In U.S. made by Westinghouse, Combustion Engineering ans Babcock and Wilcox.
QUAD	One quintillion BTUs, 10^{18} BTUs.
Radioactivity	The property of some unstable atoms to disintegrate spontaneously.
Rasmussen Report	A 1975 AEC report (WASH-1400) on nuclear accidents prepared under Prof. Norman Rasmussen of M.I.T.
Scram	Term applied to the sudden shutdown of a nuclear reactor by insertion of safety control rods.
Secondary Recovery	A technique to extract additional oil or gas from a formation once reservoir pressure is inadequate to sustain a flow.
Solar Cell	A photovoltaic device for conversion of solar energy into electrical energy.

Solar Energy	Energy from the sun.
Spent Fuel	Fuel that has been discharged from a nuclear reactor for reprocessing.
Strip Mining	Production of coal from deposits by stripping rock and soil overburdens.
Tar Sand	Sand formations containing oil.
Tcf	Trillion cubic feet.
Tertiary Recovery	Technique using heat and/or chemicals to recover additional oil from reservoirs after use of secondary recovery methods.
Thermal Pollution	Discharge of heated water into lakes, rivers and oceans.
Thermonuclear Reactor	Fusion of light elements such as hydrogen by heat.
Tokamak	A magnetic device for controlled release of thermonuclear energy.
U_3O_8	Uranium oxide, a yellow-colored ore used in nuclear power.
U-235	Uranium-235, the fissionable isotope of uranium, constitutes o.7% of natural uranium.
U-238	Uranium-238, the heavy isotope of uranium.
Ultraviolet Radiation	Radiant energy of very short wave-length.
USGS	United States Geological Survey.
Watt	Unit of electric power.
X-Ray	Radiation from an X-ray tube.

SELECTED BIBLIOGRAPHY

American Gas Association, Gas Supply Committee, *Gas Supply Review,* Volumes 1-6 (Virginia, 1977)

American Gas Association, Gas Utility Industry, *Projections to 1990*, (Virginia, 1973)

American Gas Association, *Gas Data Book-Brief Excerpts from Gas Facts 1971 Data,* (Virginia, 1972)

American Gas Association, *Elements of an Effective Energy Program,* (Virginia, N.D.)

American Petroleum Institute, *One Answer to the Energy Crisis,* (Washington, D.C., 1972)

American Petroleum Institute, *Publications and Materials,* (Washington, D.C., N.D.)

Baedeker, Karl, *The United States: A Handbook for Travellers — 1893*, (New York: Da Capo Press reprint, 1971)

Beckmann, Petr, *Making Plutonium A Soviet Monopoly,* (Nashville, Tenn: United States Industrial Council, N.D.)

Beckmann, Petr, *Nuclear Proliferation: How to Blunder into Promoting It,* (Boulder, Colorado: The Golem Press, 1977)

Beckmann Petr, *Pages From U.S. Energy History,* (Boulder, Colorado: The Golem Press, 1978)

Beckmann, Petr, *The Health Hazards of Not Going Nuclear*, (Boulder Colorado: The Golem Press, 1976)

Benedict, Manson, *The Case for Nuclear Power*, (Washington, D.C., National Academy of Engineering, 1976)

Boesch, Donald F.; Hershner, Carl H.; Milgram, Jerome H. *Oil Spills and the Marine Environment*, (Cambridge, Mass.: Ballinger Publishing Company, N.D.)

California Energy Resources Conservation and Development Commission, *Nuclear Fuel Reprocessing and High Level Waste Disposal: An Interim Report*, 1977

Edison Electric Institute, *Economic Growth in the Future*, (New York, N.Y.: 1976)

Edison Electric Institute, *EEI Pocketbook of Electric Utility Industry Statistics*, (New York, N.Y.: 1972)

Energy Conservation Research, *Our Energy-Problems & Solutions*, (Malvern, Pa.: 1977)

Energy Research and Development Administration, *Coal In Our Energy Future*, (Washington, 1977)

Energy Research and Development Administration, Energy Research and Development Administration Bicentennial Project, *Energy History of the United States 1776-1976*, (Washington, D.C.: 1976)

Energy Research and Development Administration, *So What's New?*, (Washington, D.C.: 1975)

Engler, Robert, *The Politics of Oil*, (New York: The Macmillian Company, 1961)

Federal Energy Administration, *A Federal Role in Natural Gas Distribution Rate Structure Reform*, (Washington, D.C.: National Energy Conservation Programs, 1977)

Federal Energy Administration, *Project Independence*, (Springfield, Virginia: National Technical Information Service, 1974)

Federal Energy Administration, *Technical Reports of the Federal Energy Administration,* (Washington, D.C.: National Energy Information Center, 1977)

Federal Power Commission, *National Gas Supply and Demand 1971-1990,* (Washington, D.C.: Bureau of Natural Gas, 1972)

Federal Power Commission, *1972 Annual Report,* (Washington, D.C.: U.S. Government Printing Office, 1972)

Final Report Corporate Environment Study, *The Business Environment in 1975-1985,* (New York: The Hudson Institute, Inc. 1974)

Final Report of the National Commission on Materials Policy, *Material Needs and the Environment Today and Tomorrow,* (Washington, D.C.: U.S. Government Printing Office, 1973)

Forbes, Ian A., *Energy Strategy: Not What But How,* (Framingham, Mass.: The Energy Research Group, Inc., 1977)

Grayson, Melvin J. & Shepard, Thomas R., Jr., *The Disaster Lobby,* (Chicago: Follett Publishing Company, 1973)

Hayes, Denis, *Energy: The Case for Conservation* (Washington, D.C.: Worldwatch Institute, 1976)

Hayek, Friedrich A., *The Road to Serfdom,* (Chicago: University of Chicago Press, 1944)

Hayes, Denis, *Nuclear Power: The Fifth Horseman,* (Washington, D.C.: Worldwatch Institute, 1976)

Hayes, Denis, *Energy: The Solar Prospect,* (Washington, D.C.: Worldwatch Institute)

Hoenes, G.R. & Soldat, J.K., *Age-Specific Radiation Dose Commitment Factors for a One-Year Chronic Intake,* (Springfield, Virginia: National Technical Information Service, 1977)

Ikard, Frank N., (Remarks by) *A Perspective on the Nation's Energy Supply Problems,* (Washington, D.C.: American Petroleum Institute, 1973)

Katz, Howard S., "The Case for Political Action," *Reason,* (April, 1975)

Kahn, Herman, *The Next 200 Years,* (New York: William Morrow, 1976)

Kane, Daniel W. *Council on Energy Independence,* (Chicago)

Kennedy, Edwin L., *Fuel for the Future: The Economics of Energy Supply,* (Washington, D.C.: American Petroleum Institute, 1972)

Lapp, Ralph E, *America's Energy* (Greenwich, Conn: Reddy Communications, Inc. 1976)

MacAvoy, Paul W., *Price Information In Natural Gas Fields* (New Haven and London: Yale University Press, 1962)

MacAvoy, Paul W. & Sloss, James, *Regulation of Transport Innovation,* (New York: Random House, Inc., 1967)

Males, Rene H., and Richels, Richard G. *Economic Value of the Breeder Technology: A Comment on the Ford-Mitre Study* (April 12, 1977)

Medvin, Norman; Lav, Iris J, & Ruttenburg, Stanley H., *The Energy Cartel,* (New York, N.Y.: Marine Engineers' Beneficial Association, 1975)

Ministry of Natural Resources, *Ontario Mineral Review* (Ontario: Ministry of Natural Resources, N.D.)

Mitchell, Edward J., *The Energy Dilemma: Which Way Out?* (Washington, D.C.: American Enterprize Institute, 1975)

National Association of Manufacturers (NAM), *A National Energy Policy,* (New York, N.Y.: N.D.)

National Coal Association, *Bituminous Coal, Facts 1972,* (Washington, D.C. 1972)

National Coal Association, *Coal Facts, 1974-1975,* (Washington, D.C.: 1975)

National Coal Association, *Coal Makes the Difference*, (Washington, D.C.: 1973)

National Coal Association, *Toward Responsible Energy Policies*, (Arlington, Virginia: American Gas Association, Washington, D.C.; American Petroleum Institute, New York, N.Y.: Atomic Industrial Forum, Inc, New York, N.Y.: Edison Electric Institute, Washington, D.C.: 1973)

National Petroleum Council, *Guide to National Petroleum Council Report on United States Energy Outlook* (Washington, D.C.: 1972)

Page, Earl M., *We Did Not Almost Lose Detroit*, (Detroit: Detroit Edison, 1976)

Peterson, William H., *The Quest of Governmental Oil Import Restrictions*, (American Enterprise Institute, 1959)

Price, D.L., *Stability in the Gulf: The Oil Revolution* (Great Britain: The Eastern Press, Ltd., London and Reading, 1976)

Report of the Nuclear Energy Policy Group, *Nuclear Power Issues and Choices*, (Cambridge, Mass.: Ballinger Publishing Company, 1977)

Rocks, Lawrence & Runyon, Richard P., *The Energy Crisis*, (New York: Crown Publishers, Inc. 1972)

Sobel, Lester A., (edited by) *Energy Crisis 1945-75*, (New York, N.Y.: Facts on File, Inc., 1974)

Teller, Edward, *Energy: A Plan for Action*, (New York, N.Y.: Commission on Critical Choices for Americans, N.D.)

The Hudson Institute, Vol. I, *The Business Environment In 1975-1985*, (New York, N.Y.: 1974)

The President's Materials Policy Commission, *Resources for Freedom: Foundations for Growth and Security*, (Washington, D C.: 1952)

The Edison Electric Institute, *The Transitional Storm*, (New York, N.Y.: 1977)

U.S. Congress, National Commission on Materials Policy, *Material Needs and the Environment Today and Tomorrow*, (Washington, D.C.: 1973)

U.S. House of Representatives, Hearing before the Committee on Ways and Means, *Tax Aspects of President Carter's Energy Program*, 95th Congress, 1st Session (Washington, D.C.: 1977)

U.S. Senate, Permanent Subcommittee on Investigations of the Committee on Government Operations, *Materials Shortages*, 94th Congress, 1st Session, (Washington, D.C.: 1975)

U.S. Energy Policies — An Agenda for Research, (Maryland: Resources for the Future, Inc., 1975)

Wilson, Richard, *Bulletin of the Atomic Scientist*, (May, 1972)

INDEX

Books In Focus believes that the following Newsletters (some of which are helping to spread the viewpoint of this book) will be of special interest to readers:

Value-Action Advisory Service
3838 48th Avenue, NE
Seattle, Washington 98105
Call the *Voice of Gold* (206) 525-7500

North American Client Advisory
100 West Washington Street
Phoenix, Arizona 85003

The Forecaster
19623 Ventura Boulevard
Tarzana, California 91356

Gold Newsletter
8422 Oak Street
New Orleans, Louisiana 70118

Daily News Digest
P.O. Box 39027
Phoenix, Arizona 85069

Pattersen Strategy Letter
P.O. Box 37432
Cincinnati Ohio 45237

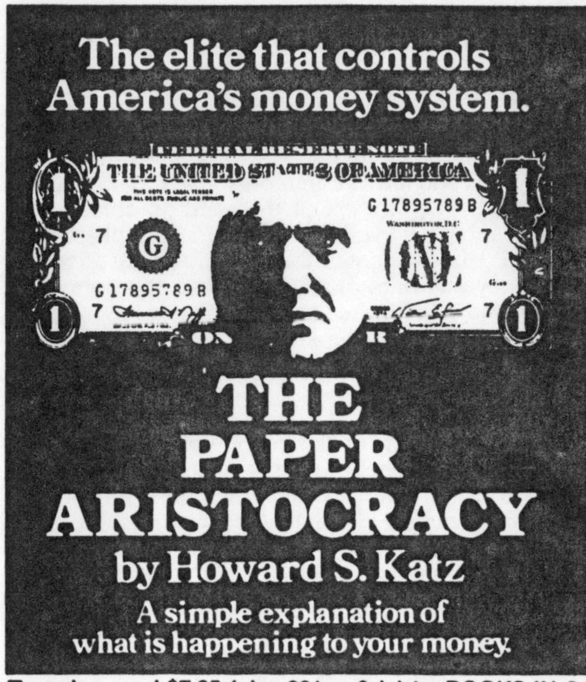

Katz has written a new classic as important to the development of economic thought as Adam Smith's *Wealth of Nations*, or Marx's *Das Kapital*. For in this book Katz carefully documents the development of a powerful elite, which through the mechanism of paper money, has come to control the economies of Nations. This group, which Katz calls the paper aristocracy, reaps unearned benefits from this control at the expense of the people. For the first time ever, economics is not viewed as a cold science, but is evaluated within a moral framework. What's more, Katz does it in clear language easily understood by the layman or student.

The Paper Aristocracy represents a major revolution in economic thinking. Howard Katz deserves the praise and thanks of all honest men. Order his book now.

"Your presentation is about the most penetrating and stimulating of any I have read in years. You are absolutely right—paper money inflation makes the poor poorer and the rich richer."

—ELGIN GROSECLOSE, PH.D.,
Institute for Monetary Research, Inc.,
Washington, D.C.

To order, send $7.95 (plus 80¢ p. & h.) to: BOOKS IN FOCUS. 30 day return privilege.

Another classic by Howard S. Katz.

THE WARMONGERS

They started World War I.
They engineered World War II.
They are moving the world toward war now!

Who they are, how they have operated, and above all why they are warmongers, is carefully documented in this brilliant analysis.

THE WARMONGERS re-examines history and discovers a distressing link between the creation of paper money and major wars. Detailed facts expose a fascinating and frightening intrigue in which bankers, big business and government create wars to increase their power and wealth.

THE WARMONGERS analyzes the present and uncovers a consistent trail of actions on the highest level, designed to lead the U.S. into major war.

THE WARMONGERS projects the future, convincingly predicting that America and China will be fighting Russia and Japan; tells what major policy shifts must occur; and explains why 1981 and 1985 are the most likely years for the U.S. to be brought into the war.

THE WARMONGERS is a non-fiction book.

To order, send $11.95 (plus 80¢ p. & h.) to: BOOKS IN FOCUS. 30 day return privilege.

Conflict of Minds

Changing Power Dispositions in South Africa

by Jordan K. Ngubane

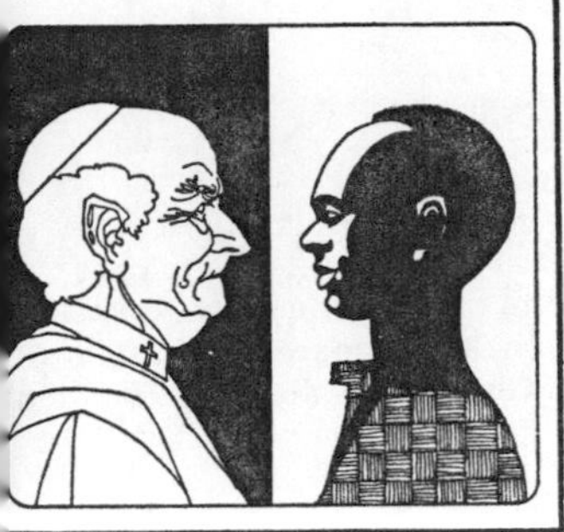

Conflict of Minds is a brilliant redefinition of the South African Crisis in terms of two conflicting philosophies of life, rather than a racial struggle.

Ngubane shows how the Sudic philosophy of Africa, and attitude toward the person evolved from ancient Egyptian theology, being passed on from generation to generation through the oral tradition; and he traces it to present day Zulu philosophy.

Conflict of Minds is the first written presentation of this philosophy ever, and therefore represents a major cultural achievement. It is a clear demonstration that there is a unifying world view throughout Africa, and that contrary to popular belief this view is in respects actually in advance of the Judeo-Christian structure. This has implications for that structure which cannot be ignored.

ABOUT THE AUTHOR:

Conflict of Minds represents a labor of 10 years for Mr. Jordan K. Ngubane. During this period he has been in exile from his home and people in the Zulu areas of South Africa. Living in Washington, D.C. and lecturing at Howard University, Jordan has maintained an objective attitude and has remained very much in touch with Black leadership within South Africa.

This is Jordan Ngubane's fifth book. He was the editor of a South African Black newspaper for eight years and was the correspondent for Mahatma Ghandi's newspaper. He has written 10 articles for the McGraw-Hill New Encyclopedia of World Biographies.

To order send $10.95(plus 80¢ p.&h.) to Books In Focus. 30 day return privilege.

"It is by far the best and most comprehensive statement for liberty I have come across."

Leon Louw
Founder and Executive Director
Free Market Foundation

In Search of Liberty

FRED MACASKILL

IN SEARCH OF LIBERTY advances the concept that the more liberty a society enjoys, the stronger it will be, with greater material and moral benefits available to its individual members.

THE PRINCIPLES AND CONCLUSIONS of *In Search of Liberty* are based on the solid bedrock of individual rights. Taking nothing for granted, the author proceeds from the most basic right - the right to live. Then, considering more complex derivative rights, *In Search of Liberty* examines the most pressing issue of our time - the proper role of government in the life of man. Statism is shown to be destructive to individual rights; the free market is demonstrated to be an approach consistent with man's nature. These free market conclusions are underscored by the fact that author Macaskill lives in the highly authoritarian South African society. The concluding chapter appropriately shows how the individual rights approach (without regard to color or nationality) is the best and possibly the only hope in the South African situation.

"It's no exaggeration to say that the future of Western Civilization depends on ideas set forth in this book."

James U. Blanchard
Chairman, National Committee
for Monetary Reform

To order, Send $8.95 (+80¢ P. & H.) to: BOOKS IN FOCUS 30 Day Return Privelege

SPECIAL DISCOUNTS TO READERS

The following quantity discounts are available to readers who want to help spread the message of this book. (The same discount applies to all books available through Books In Focus).

10 copies 20%
25 copies 30%
50 copies 40%
100 or more 45%

Add 25¢ per book for postage and handling. New York residents add sales tax also. Mail your order and check to: BOOKS IN FOCUS, P.O. Box 3481, Grand Central Station, New York, N.Y. 10017.

To Order Books

Send The Following Information:
(You can send this sheet or photocopy it)

Book Title(s)	Quantity	Cost

Total quantity:

Name: ____________________
Address: ____________________
City: __________ State: ______
ZIP: _____

Total Cost: ______
Minus % Discount: ______
Plus Postage: ______
Total Due: (check included) ======

Send your personal check or money order payable to:

BOOKS IN FOCUS, INC.
P.O. BOX 3481
GRAND CENTRAL STATION, New York 10017

If you have a question, phone (212) 490-0334